# First Steps in Proven Geometry

## For the Upper Elementary Grades

# First Steps in Proven Geometry

## For the Upper Elementary Grades

### Geometry Lessons in the Waldorf School
### Volume III

by

Ernst Schuberth

Printed with support from the Waldorf Curriculum Fund

Published by:
Waldorf Publications at the
Research Institute for Waldorf Education
38 Main Street
Chatham NY 12037

Title: *First Steps in Proven Geometry:*
   *For the Upper Elementary Grades*
Author: Ernst Schuberth
Translator: Nina Kuettel
Editor: David Mitchell
Proofreader: Ann Erwin
Cover: Hallie Wootan
© 2009 by AWSNA
ISBN # 978-1-888365-79-5

# Table of Contents

**Foreword** ............................................................. 7

**Introduction** ........................................................ 9

**Angles in Regular Polygons** ............................. 12

**Triangles** ............................................................ 23

    The Equilateral Triangle ................................... 23

    The Tetrahedron ............................................... 28

    The Regular Hexagon in a Circle ................... 29

    The Isosceles Triangle ..................................... 30

    The Right Angle Triangle ................................. 33

    The Scalene Triangle ....................................... 34

    The Sum of the Interior Angles in a Triangle ... 36

    Theorem of the Sum of the Angles in an $n$-Angle Polygon ... 38

    The Exterior Angles of a Triangle ................... 40

**The Right Triangle and the Circle: The Theorem of Thales**   44

**The Rules of the Right Triangle** ........................ 54

    The Theorem of Pythagoras ............................ 54

    Supplementary Proof (the Indian Proof) ......... 61

    The Reverse of the Pythagorean Theorem ... 64

    Addendum to the Pythagorean Theorem ....... 71

## The Principles of Congruence and Basic Assignments Using the Triangle .......... 73

    The Principles of Congruence .......... 73

    Basic Assignments Using the Triangle .......... 74

## The Important Lines of a Triangle .......... 80

    The Perpendicular Bisector of a Triangle .......... 80

    The Circum-Circle .......... 81

    Heights of a Triangle .......... 84

    The Angle Bisectors in a Triangle .......... 91

    The Inscribed-Circle .......... 92

    The Adjacent Circles of a Triangle .......... 94

## The Platonic Solids .......... 95

    The Language of Geometric Forms .......... 97

    A Beginning of Conceptual Understanding .......... 98

    There Are Only Five Regular Solids .......... 99

    Constructing the Platonic Solids .......... 101

    Simple Projections .......... 104

## Teaching Projection and Shadow Drawing in the Upper Elementary Grades .......... 110

## Geometry and Mineralogy .......... 123

## Outlook .......... 125

## Endnotes .......... 127

## Bibliography .......... 132

# Foreword

This third volume in the series titled *Geometry Lessons in the Waldorf Schools*, following the volumes *Form Drawing* (grades one through four), and *Comparative Geometry* (grades four and five), leads into *First Steps in Proven Geometry*. Like the other books it addresses, in the first instance, the class teachers of Waldorf schools and offers suggestions and inspiration for their geometry lessons. I remember gratefully classes in geometry when I was a student of the Waldorf schools in Hannover and later in Wuppertal where I felt the beauty and truth of geometric figures and theorems. May this small booklet help young people to have similar experiences.

An interesting question is—if the computer can do it much faster and more precisely, should we continue to produce wonderful colored drawings on our own? The answer depends on our goal—do we want to produce something perfect (whatever that means) or do we want do develop the student's soul, his or her joy in producing something? Watch the young people doing it and you will see the effect. The development of aesthetic feeling and fine motor skills is a more general aspect of such a work. This remark does not mean Information Technology (IT) should never be used. But it seems to be more and more important to let people work with IT that develops creativity above technological raster.

I owe gratefulness to many people who helped me to write this book—my colleagues, my unforgotten secretary Monika Feles-Baumann, who died this year from a painful sickness, Klaus Volkert, who as a colleague read the text carefully (all remaining mistakes are my responsibility—and, sadly, always some remain—Elisabeth Wannert, who shared some wonderful drawings, and many others including all the students in my courses.

The English version has been made possible only because my wonderful friend and colleague David Mitchell as director of the AWSNA publications has been willing to care for it. His unbelievable patience and endurance with my schedule gave the project the chance to survive. I also express thanks to his publication team, Nina Kuettel as translator and Ann Erwin as proofreader. Thank you all from the bottom of my heart.

– Ernst Schuberth
Mannheim, Germany
December 2008

# Introduction

In the sixth grade the children pass into an important area of their development in which there is a new desire awakened for causal thinking. This need should be met by all the different subjects studied in school. At previous age levels, causes and effects were experienced and used, but, in general, it is only now that the desire to recognize logical connections comes to the fore. Geometry is especially suited to fulfill this desire. At the same time, geometry has a healthy effect on a certain aspect of fixated thinking that also appears at this age. Those who believe they are right must be able to bring forth good reasons, and without quarreling! Even though all of life's questions cannot be handled in a geometric sense, the study of geometry can teach us that there are areas of thinking in which truth can be well differentiated.

Preadolescence, with its strong drive for independence in thinking and aesthetic judgment, appears as a pre-illumination of the far-reaching transformation that actual adolescence brings. The Waldorf School curriculum endeavors to fulfill this need while still under the direction of the class teacher, without premature anticipation of actual adolescence. This special situation should be taken into consideration if, in the following pages, some things appear to be not very systematically and theoretically complete. There are more paths of exploration in the area of theoretic relationships between the geometric forms than recent completed science.

This book builds upon work done previously in the fourth and fifth grades; especially basic geometric construction (See *Geometry Lessons in the Waldorf Schools, Volume II*).[1] It is also expected that the geometry block in the sixth grade is preceded by an algebra block in which the formulae for interest rates and other principles have been taught.[2] When first looking at the polygon, this is the intention: It is about finding general principles that encompass each separate case and lead to a pure mental understanding.

Even the lessons in sentence construction that are begun in sixth grade grammar studies—with the various parts of the sentence such as main clauses, subordinate clauses, and looking at the function of each conjunctive word—is completely connected to this. Imaginatively speaking, one could say: A "Jupiter impulse" can be cultivated in twelve-year-old children through the understanding of theoretical relationships.[3]

This book contains more than can normally be covered during one year. So, it will be necessary to pick and choose in such a way that the whole represents meaningful connections. The main goal is not the amount of material covered, but rather, the most important impulse lies in experiencing the joy of learning about pure geometric forms that children in this age, with their strong intellect, so enthusiastically display like no other time in their later lives.

However, many children experience not only joy, but also frustration because a sought-after solution to a problem can not be found. One can confidently point out to the children that abilities are best cultivated when a solution is *not yet* in sight, but we are diligently searching for it. The different construction assignments using triangles are especially well-suited for this purpose. We, as teachers, should not be shy about assigning things which the children find very difficult. Many children shy away from this because their frustration tolerance is often too low to persevere until they have found the solution. In these cases we must be encouraging. Intellectual work is just as difficult as physical work and it also strengthens us in the same way. Teachers can show the children, step by step, what systematic work on an intellectual task looks like. Above all, the children must learn to ask systematic questions: What is given? What is being asked? What can I do with what is given? What principles do I already know that have to do with the question? In this way, after the teacher has carefully worked through the problem with the children, they will be asked to find and describe other problems themselves.

In Waldorf schools, rightfully so, we try to cultivate a sense of beauty along with searching for insight (truth); this can be schooled through the aesthetic language of geometric forms. The well-known master of mathematics lessons in the Waldorf school, Hermann von Baravalle, gave many helpful suggestions in his classic work *Geometry, the Language of Form*.[4] As far as it is possible within the limited space here, there are examples given from the language of form. But many beautiful things can be easily discovered on one's own. However, one should always differentiate: When is it a merely decorative form and when is a higher mathematical principle to be seen?

Many colleagues in the Waldorf schools wish to build their geometry lessons upon the foundation of classic *projective geometry* such as it is presented on a higher level in the works of Louis Locher-Ernst, for example.[5] Although basic geometry can be developed out of projective geometry, I believe that the educational significance of classic Euclidean geometry is in no way obsolete, and what follows in this book takes, in many ways, traditional paths. Those who are knowledgeable will find many things in the background that can lead to a projective way of thinking.

# Angles in Regular Polygons

In the fourth grade we described how elliptical movement was created gradually out of circular movement and also how the forms of regular triangles and squares can be attained.[6] John physically experienced the turning motion while walking a circle. We will pick things up again at this point in order to go deeper into the lawfulness of certain figures, starting with the regular polygons.

With the algebraic description of principles found in regular polygons, the basic impulses of the algebra lessons are brought up again, and, at the same time, a good utilization of those principles. It is left up to the teacher to decide how far to go into a review of the fourth grade material, whether to walk through everything again or simply recall the material to mind. The important thing is that the children are convinced that when they walk a circle they are rotating 360°. Since all the readers may not be familiar with the second volume of *Geometry Lessons in the Waldorf School,* I will take the opportunity here to reiterate something from that book. It is helpful if the following can be put before the children again in as lively a way as possible. In my class the entire process went something like this:

"John, come to the front of the class, please!" John, who is strong and very adept at solving technical problems but less so in artistic and scientific endeavors, comes to the front. "Make a 360° turn!" John turns around once on his own axis. On command, he executes various angles and thereby shows that he has mastered angle measurement.

"Walk one complete circle!" John walks a circle. "Did you also turn while walking?"

"No!" So we begin again.

"Make another 360° turn and describe what you see!" John makes the turn and tells the class what he sees while turning. He sees the door, the table, the window, and the class once again.

"Now, walk a circle again!" John looks at the floor and good-naturedly walks another circle. "Did you turn while walking this time?"

"No!"

"All right, walk it again and tell us what you see in front of you!" John starts walking and says: "The class, the door, the table, the window, the class."

"Did you turn while walking?"

"Yes." But he is still not completely convinced. So, the class describes how they see him. First, he turns once more on his axis. The class sees him full-front, then his right shoulder, his back, his left shoulder, and finally his full-front again. Then he walks a circle once again and the class sees him the same way from different sides just as before.

Whether he wants to or not, John must accept that he is turning while walking in a circle. But I can see that he is still not absolutely convinced, so I try another tack. "What is the difference if one turns or goes straight ahead?" (This introduces two terms in projective geometry: *Rotation* and *straight-line movement*.)[7] Finally, someone says: "When you turn you get dizzy and when you walk straight ahead you don't." If one is rotating while walking a circle, then one must become dizzy. Robert, who has a purely sanguine temperament, immediately offers to test out this hypothesis. As quickly as possible, he walks a small circular path. After five rounds he is completely dizzy. John, who is of a more phlegmatic nature, tries it also, but a little more slowly. It is true; one becomes dizzy while walking in a circle so one is indeed turning! After this has been perceived in the depths of the physical organization, gradually, one is truly convinced that something about the whole thing is correct.

At this time, the teacher draws a picture on the blackboard of the circular path that has been walked:

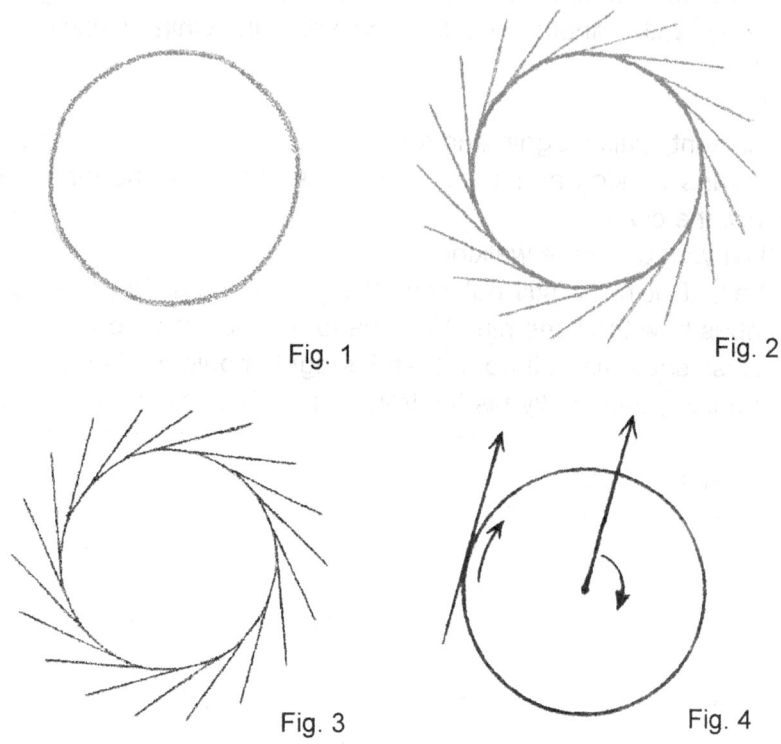

*Figs. 1–4: Rotation and Straight-forward Movement along a Circular Path*

John walks the circle once again and another child, using a stick (tangent) and referencing the figures on the blackboard, pinpoints each location and its line of sight as John is walking. If one walks forward in a circle one time, then one has rotated exactly 360°. John takes his seat, a little tired, but happy at a job well done.

Rotation can also be shown from the middle of the circle. To do this, we have one child stand in the middle with his or her gaze *always in the same direction* as that of the child walking in a circle, but *not* looking at the child who is walking.

After the resurrection of these earlier experiences with rotating and straight-forward movement while walking in a circle we continue: How is it now when, instead of a circle, we walk an oval, i.e. ellipse?[7] Katherine comes forward and walks it. The longer the oval becomes, the clearer is the rhythmic alternation between strongly emphasizing either the forward motion or the turning motion. Depending upon the curvature, either straight-forward motion (places with little curvature) or turning motion (places with strong curvature) is emphasized. On the whole, one still rotates 360° when walking an ellipse the same as when walking a circle.

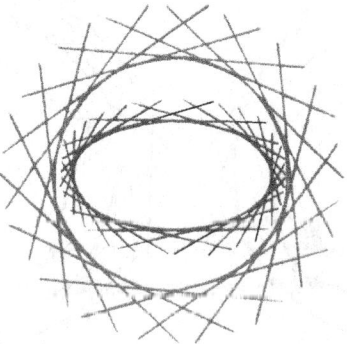

*Fig. 5: Rotation and Forward Movement of an Ellipse*

The teacher now asks Joanna to walk a "three-corner oval." With every completed round there is a triple rhythmic alternation between emphasis on the turning motion or the straight-forward motion. But the whole angle of rotation remains 360°.

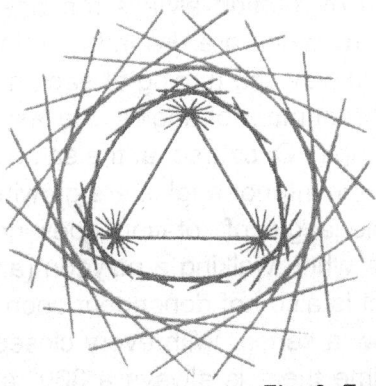

*Fig. 6: From a Circle to a Triangle*

Now we will let the alternation become increasingly abrupt until the turning rotation and straight-forward rotation are completely separated. We have come to the equilateral triangle: Along the sides we have only straight-forward motion and in the corners only rotation. In this figure the two basic forms of movement "break apart." However, with completion of the figure, the rotation angle remains 360°.

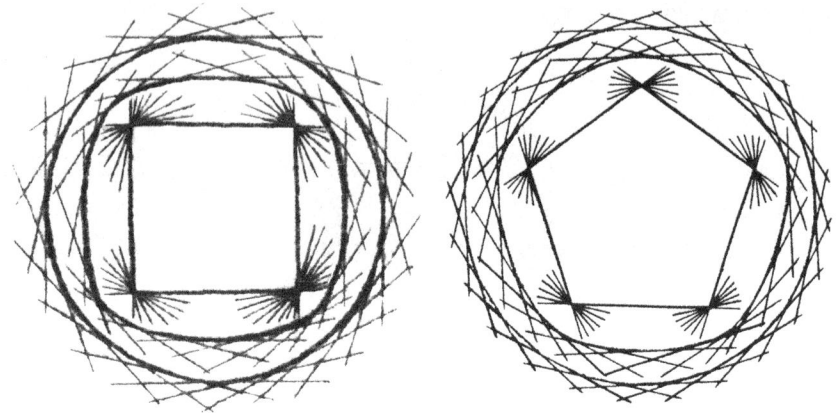

*Figs. 7 and 8: The Rotation in a Rectangle and Pentagon*

Before we pursue further the angle of a triangle, let us widen the perimeters of our question just a little. First, we prove to ourselves that the same process can lead also to a square, pentagon, hexagon, etc. The whole angle of rotation always remains 360°. The only difference is that the rotations are divided over increasingly more places (angles). In this way, the turning at each angle will become sharper. The larger the number of angles, the less one turns when rotating each single angle. Of course, at the same time the exterior angles are shrinking, the interior angles are growing.

Just as the whole angle of rotation does not depend upon the number of angles when walking a polygon (and here one can introduce this term), it is also not dependent upon the regularity of the form in too narrow a sense. With every closed curve that one "goes around" one time there is always a 360° angle of rotation. Something new arises when one makes "indentations." Without our going too much into the technicalities, the closed, convex curves and polygon forms are differentiated from the non-convex. Someone points out that one also turns 360° when walking a "bean form" since one comes back to the starting point. Between A and B, however, one turns the other way round. It must be subtracted from the rest of the rotation angles.

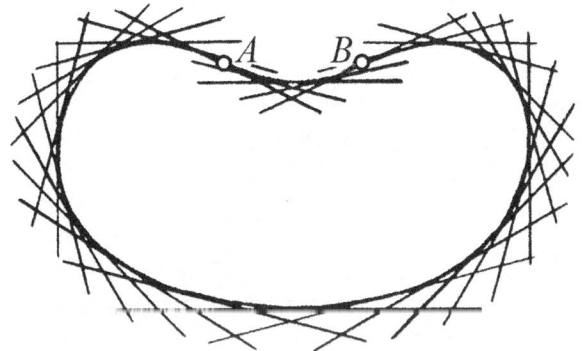

*Fig. 9: Rotation When Walking a Bean Form*

When a child slowly walks the form and the class pays attention to which side is toward the class, the various turning directions and, with that, the *turning points* can be observed. However, these observations are merely a small digression and soon we must turn back to the main subject.

When we walk an equilateral triangle we rotate 360°. We draw the figure once again and put in the rotation angles. John comes forward again, puts some kind of rod or stick under his arm that is always pointing the same direction as his nose, and demonstrates to everyone where, and at what angles, he turns. These angles are called the *exterior angles* of a triangle. One then names the *interior angles* and shows them on the blackboard.

Now we can think about the size of the individual angles and write them on the figure. The 360° rotation is divided equally among the three angles. That means at every angle one turns 360°÷ 3 = 120°. That is the size of one exterior angle. Every interior angle supplements its exterior angle for a total of 180°. So the interior angle is 60°, and all the interior angles together total 180°.

Sum of the Exterior Angles $\quad S_{EA} \quad\quad\quad\quad\quad = 360°$
One Exterior Angle $\quad\quad\quad W_{EA} = 360° \div 3 = 120°$
One Interior Angle $\quad\quad\quad\; W_{IA} = 180° - 120° = 60°$
Sum of the Interior Angles $\quad S_{IA} = \quad 3 \times 60° = 180°$

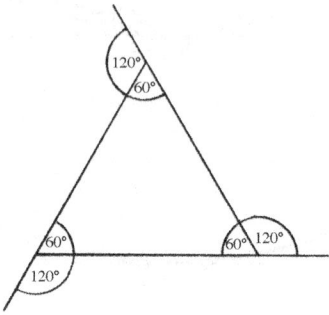

*Fig. 10: The Exterior and Interior Angles of an Equilateral Triangle*

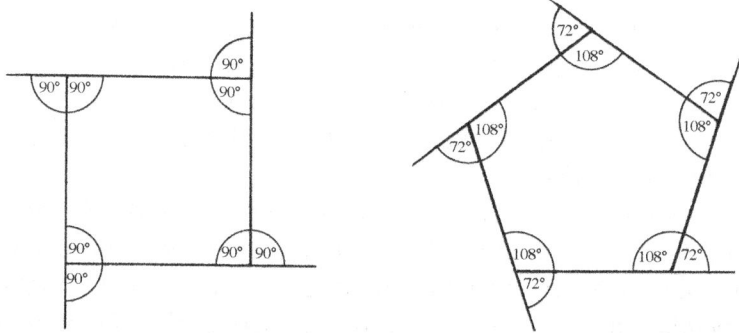

*Figs. 11 and 12: The Sum of the Angles in a Square and a Pentagon*

Let us look at the angles of a square and pentagon. The 360° is divided into four and five equal parts respectively. This makes

$$360° \div 4 = 90°$$

for one exterior angle of a square and

$$360° \div 5 = 72$$

for one exterior angle of a pentagon. The students can calculate one of these two examples as they choose, either individually or in small groups.

In a square, the sum of the interior and exterior angles are the same (that is, 360°). In a pentagon the sum of the interior angles begins to be more: 5 x 108° = 540°. We write the results in a table as follows:

| Number of Angles | A | Sum of Exterior Angles $S_{EA}$ | Size of Exterior Angles $W_{EA}$ | Size of Interior Angles $W_{IA}$ | Sum of Interior Angles $S_{IA}$ |
|---|---|---|---|---|---|
| Triangle | 3 | 360° | 120° | 60° | 180° |
| Square | 4 | 360° | 90° | 90° | 360° |
| Pentagon | 5 | 360° | 72° | 108° | 540° |

**Homework Assignment:** Calculate the sizes and sums of the exterior and interior angles for a hexagon and a decagon. Those who wish may fill out the table to include a hexagon and a decagon.

The important things will be copied into the subject book. Then a short preview for the next day's lesson will be given. The last part of the lesson is the narrative whereby stories from history appropriate for the sixth grade are told.

**Remarks:** Besides the rhythmic development of each main lesson which usually begins at 8:00AM and lasts between 90 and 120 minutes, the development from day to day is very important. In this way, when one is beginning a new subject, on the first day concrete examples can remain in the foreground of the lesson with their important points being brought back to memory during a review at the end of the lesson period. After a night's sleep, we are psychologically in a different relationship to our experiences of the day before. Unconscious mental forces work through the material again during sleep, which provides a changed disposition to learning the next morning. The result is a more thoughtful penetration of the material and a deeper understanding. More details on this subject are written about in anthroposophical literature[9] and confirmed by modern brain research.

The next morning the lesson should not begin with the often asked question: What did we learn yesterday? But rather, it could go something like this: "Yesterday, we talked about the angles and sums of angles in a regular triangle, square, and pentagon. The calculation process was very similar in all the examples. Your homework assignment was to calculate other figures using the same process. Who among you can describe the calculation process, the *principle* that applies to every *polygon*?"

Mark, a child of sanguine temperament who is often the first to ask questions, raises his hand: "With a triangle, for example..."

He is already interrupted: "Yes, we talked about the triangle yesterday. Now we are looking for the calculation process, the principle that applies to all regular polygons. Who can describe it?" Mark is a little unhappy because he knows how it should go but he just can't verbalize it quite yet. Since he can't describe the principle without the use of an example, Thomas raises his hand: "Let's say we have a square..."

"Thomas, we also talked about the square yesterday. Can you describe the principle that is at work in all the calculation processes without using an example?"

The class seems a little surprised. This type of question is still very unfamiliar. Obviously, the teacher is expecting something totally new. "What does he want from us?" is the unspoken question on their minds. Finally, Gabrielle raises her hand. She is slightly older than the other children in the class. Math is not her best subject, but she is very interested in English and any kind of storytelling. She lives under difficult circumstances at home and has more responsibilities than most children of her age.

She tries to describe the calculation process: "When we walk a polygon we turn 360°. The angle of one corner can be gotten when 360° is divided by the number of angles. An interior angle supplements the angle to 180°. If one takes this angle times the number of angles, then one gets all of the (interior) angles together." Amazed silence in the class...it works. One can describe the process without using an example. The whole thing is discussed back and forth, and then we try to formulate it in writing. To represent the number of angles we use the letter A. This is what we get:

Sum of the Exterior Angles: $S_{EA} = 360°$

Size of an Exterior Angle: $W_{EA} = 360° \div A$

Size of an Interior Angle: $W_{IA} = 180° - W_{EA} = (180° - 360°) \div A$

Sum of the Interior Angles: $S_{IA} = A \cdot W_{IA} = A \cdot (180° - 360°) \div A$

If the class is far enough along in understanding algebra, then the brackets can be multiplied out. That would give:

$$S_{IA} = A \cdot (180° - 360°)$$

One can bracket the 180°, then 360° = 2 • 180°. The result is:

$$S_{IA} = (A - 2) \cdot 180°$$

This is really a surprisingly simple and beautiful principle! We proceed immediately to proving it upon the first examples:

Triangle: A = 3
Size of an Exterior Angle $W_{EA} = 360° \div 3 = 120°$
Size of an Interior Angle $W_{IA} = 180° - 120° = 60°$
Sum of the Interior Angles $S_{IA} = 3 \cdot 60° = 180°$

Summary of the calculation:

$$S_{IA} = 3 \cdot (180° - 120°) = 3 \cdot 60° = 180°$$

If we shorten the last formula it looks like this:

$$S_{IA} = (3 - 2) \cdot 180° = 1 \cdot 180° = 180°$$

We can prove the formulas for the square and pentagon in the same way. The Formula is:

*The sum of the interior angles of a regular polygon can be calculated by reducing the number of angles by two and multiplying the result by 180°.*

One can also point out that the sum of the angles in a regular triangle is 180°, and that by raising the number of angles by one, in each case, this amount is added. This will need further explanation at a later time.[10]

Now we can quickly go over the homework assignment by filling out the table to A = 10 with the help of the formulas. The class can work in groups to do this. The homework assignment will be to calculate

$$A = 12, A = 36, \text{ and } A = 100.$$

After working on the subject books, the story part of the lesson, and a preview of the next day's lesson, the main lesson is ended for that day. Tomorrow we will take a closer look at the construction of the formula. How is it represented in the formula when the exterior angles become smaller and the interior angles become larger with an increasing number of angles?

# TRIANGLES

## The Equilateral Triangle

The equilateral triangle appears again and again in form drawing, freehand geometry, and fifth grade geometry. At first we encounter it as the simplest of all the regular polygons. Even at the beginning of a more systematic way of learning about triangles, we still start with the equilateral triangle.

Labeling: First, we call to mind the descriptions introduced in the fifth grade on the corners (points), sides, and angles of a triangle. Large Latin letters for the points, small letters for the sides (line segments), and small Greek letters for the angles, whereby $\alpha$, $\beta$, and $\gamma$ are the most commonly used. A triangle with the corners A, B, and C is written as $\triangle ABC$. A straight line can be described by giving two of its points, such as $\overline{AB}$, for example. $\overline{AB}$ describes the line segment from point A to point B (without orientation). We can represent an angle by giving three points. For this we use an angle symbol at the beginning. $\angle ABC$ describes the angle that is formed by the two lines $\overline{BC}$ and $\overline{BA}$. B is the vertex of the angle while $\overline{BC}$ and $\overline{BA}$ are the legs.

In Volume 2 we did the following problem which we repeat now. This time, however, we will require a description of the construction and the justification.[11]

Problem: Construct an equilateral triangle with side lengths of 6cm. Describe and justify the construction.

Answer: The children usually easily discover that they have to start with a segment AB length 10cm and that then $\overline{AB}$ is to be set in a compass and two intersecting circular arcs drawn from its endpoints. The point of intersection is point C of the triangle they are trying to construct. We connect C to A and B.[12]

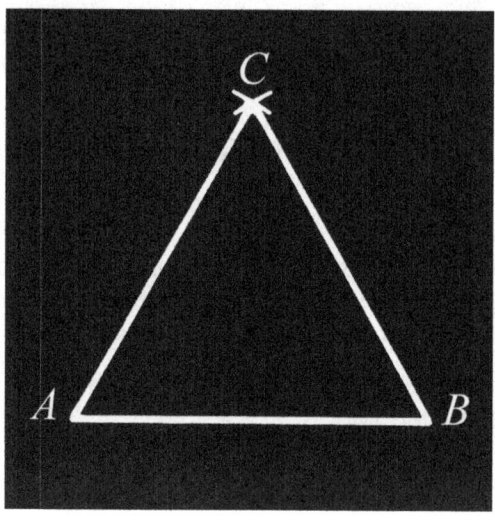

*Fig. 13: Constructing an Equilateral Triangle*

First, we need to prove to ourselves that the triangle as it is constructed has sides that are all the same length, 10cm, because all the points on a circle are the same distance from the middle point. Angle C is the same distance from angles A and B, just as angles A and B are the same distance from each other. The triangle is, therefore, equilateral.

As soon as the construction has been adequately discussed, it can be looked at from another aspect. The circular arcs used in the construction of the triangle are, of course, two parts of two complete circles. They intersect each other at *two* points, so that the exercise leads to *two* solutions. The first answer that was found appears as a fragment of the complete construction.[13]

*Fig. 14: The Two Solutions for the Complete Construction*

It is good to practice the simplified solution for *practical uses.* The necessary arcs should be visually estimated as exactly as possible. One could have a little contest on the blackboard: Who can use the smallest arcs while constructing a triangle? The given line segments $\overline{AB}$ should be presented in the most varied positions and lengths!

Starting with the result, the equilateral triangle, one sees that we could have completed the construction from *every* side. In order to express this in the constructed figure, we draw in the third circle.

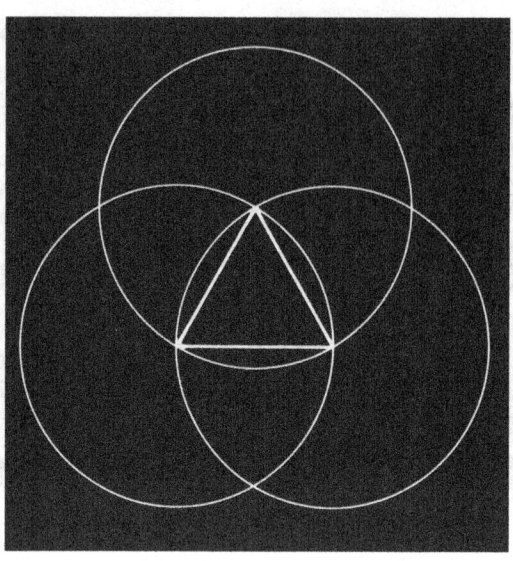

*Fig. 15: The Three Possibilities for Constructing an Equilateral Triangle*

We have now attained the *whole* that is present in the construction. This encourages a habit of thinking whereby in each exercise one asks about the associated whole. In many cases this is the only thing that facilitates a complete solution. However, as a form of thinking, its significance goes far beyond mathematics.[14]

The question arises: How can one change an equilateral triangle? One can make it larger or smaller, but one cannot change the *form*. In the following figure the steps for making an equilateral triangle larger are taken from the triangle itself. The intersection of the heights divides each height at a ratio of 1:2. The smaller segment of the height is taken as a measure for growth or reduction. In this way, the whole impression is especially harmonious. If colors are used, the inner movement and, especially, the inversion at the middle point can be emphasized.

*Fig. 16: Growing, Waning, and Inverse Equilateral Triangles*

In review, we have taken samples of the steps we went through that are described in Volume 2 at the beginning of the chapter titled *Basic Geometric Constructions*. Starting with a problem, the solution was found, described, and substantiated. Then we discussed and practiced the simplification as it applies to everyday technical uses. Finally, the associated whole was found and embedded in a beautiful figure. Further steps can be found in the following pages and be left up to the reader.

## The Tetrahedron

Four equilateral triangles of the same size can be put together to form an equilateral triangle. Are there really 180° to be measured at the places where three of the triangles come together? Yes, because every equilateral triangle has interior angles of 60° and 3 x 60° = 180°. Therefore it appears at the meeting point of a straight angle (= 180°).

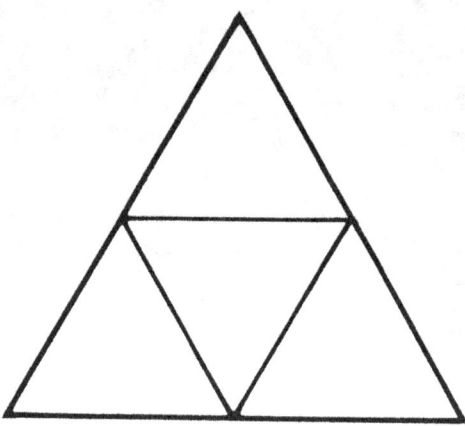

*Fig. 17: One Equilateral Triangle Formed out of Four Equilateral Triangles*

Starting with this figure, one can now take a step into the third dimension. One asks the children to form one *body* from the four triangles. The large triangle with the four equal-sided partial triangles will once again be drawn on a piece of sturdy but thin cardboard. The necessary gluestick and razor knife are discussed, and the figure—including the tabs for gluing—is cut out. The folding edges are lightly scored and the first tetrahedron is created.

A tetrahedron can be drawn on the blackboard or in a main lesson book since the subject of perspective also comes up in the sixth grade curriculum. At first I make the figures transparent, or at least partly transparent, just as they are seen in imaginative thinking. Then they are presented as solid bodies with light and shadow.

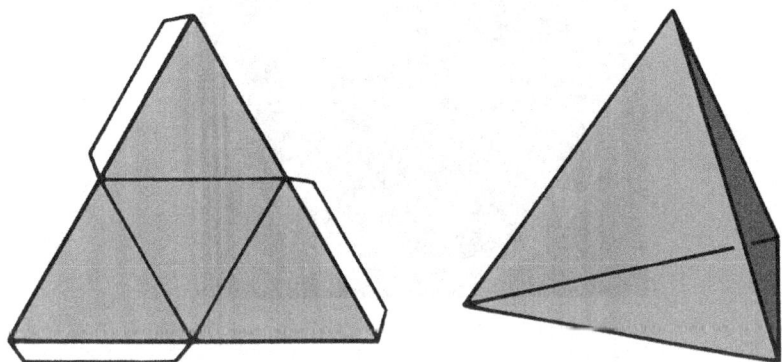

*Figs. 18 and 19: The Tetrahedron on the left represents the net with the tabs for gluing while the right is the drawing on the blackboard.*

**The Regular Hexagon in a Circle**

In association with the discussion about the equilateral triangle, now, finally, the construction of a hexagon in a circle that was done in the fifth grade can be substantiated.[15]

For that purpose we will now draw the construction carefully on the blackboard. We use a compass to carry the radius six times as a chord to its associated circle. The points attained thus are connected to form a hexagon and the three diameters are drawn in as well. It is best if the students can work in small groups to attempt to substantiate why the radius of the circle can be carried as a chord *exactly* six times. Since they have already learned that the angle of an equilateral triangle is 60°, then starting from the middle point of the circle 6 x 60° = 360° gives the complete angle. The teacher should refrain from giving the students a simple tip, thereby taking away their joy at independent discovery. The attitude of "I can find it myself" will never be as joyfully experienced as it is at the dawn of the young intellect found in this age group.

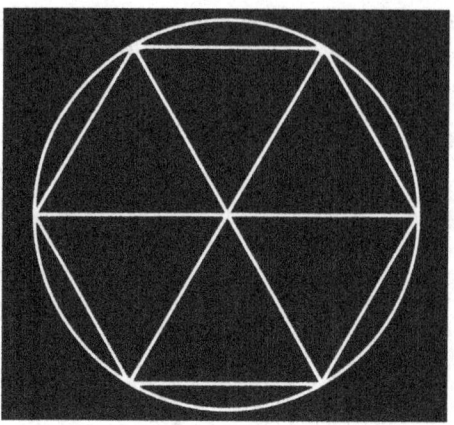

*Fig. 20: The Regular Hexagon*

**The Isosceles Triangle**

If the equilateral triangle has been adequately covered in the lessons, then we can transition in a first transformation from the equilateral triangle to an isosceles triangle.

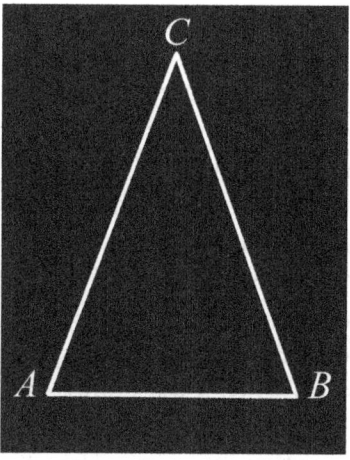

*Fig. 21: The Isosceles Triangle*

Problem: Construct triangle △ABC in which one side is 5 cm long and the other two sides are each 7 cm long. Describe and substantiate the construction (Fig. 21).

After the construction has been thoroughly discussed, the term *isosceles triangle* can be introduced. In an isosceles triangle the two equal sides are called *legs* and the third side is called the *base*. An isosceles triangle has only *one* axis of symmetry. It is the perpendicular bisector of the base. The following figures illustrate how isosceles triangles can be transformed. They are much more variable than equilateral triangles (Figs. 22 and 23).

If one stretches an isosceles triangle with the same base so that the apex slowly rises from the middle point of the base, it will become a *right angle isosceles triangle*, and later it will become an *equilateral triangle*. A right angle isosceles triangle, as a component of a square, has angles of 45°, 45°, and 90°. The sum is 180°. We are already familiar with this sum from the interior angles of an equilateral triangle. We put construction exercises on the blackboard in which the base is *not* horizontal. Then other exercises can be completed:

*Exercises for Isosceles Triangles*

1. In the middle of a sheet of paper draw a horizontal line segment that is 6–10 cm long. Construct the perpendicular bisector. From the point of intersection, draw several line segments of a determined length above and below. Construct a whole array of isosceles triangles (see Fig. 22).

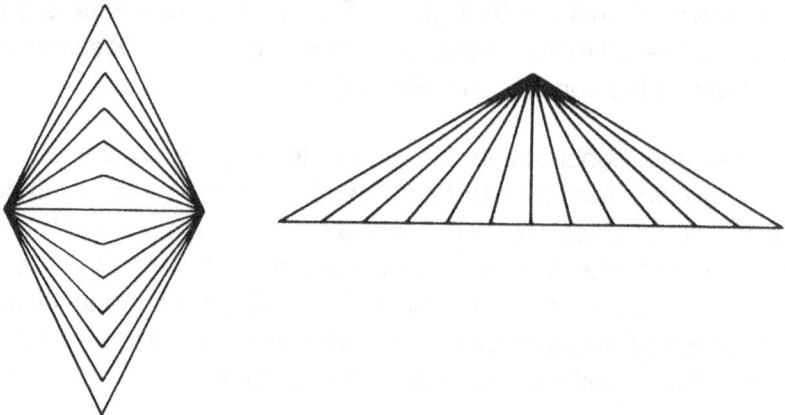

Fig. 22: Isosceles Triangles with identical bases

Fig. 23: Isosceles Triangles with the same heights

2. Make a horizontal straight line and put a point anywhere about 5cm above the line. From the point, drop the perpendicular to the line, and from that point, draw several line segments of equal lengths to the right and left. Construct an array of isosceles triangles with the same *height*. Emphasize them by using different colors.

3. Draw a circle, and starting at the top vertex (see the directions on fundamental constructions), construct the points of a dodecagon on the circle. Draw all of the isosceles triangles in which the apexes are the uppermost point of the dodecagon and the angles of the base are dodecagon points.

4. Construct a dodecagon just like in exercise three and choose the middle point as the apex for isosceles triangles whose base points on the circle line are symmetrical with the verticals. Emphasize each triangle by using a different color.

5. Construct a regular octagon in a circle. How many isosceles triangles can you locate from the points of the octagon? Label the angles with the letters A to H, and identify each isosceles triangle using its associated letters. Can you find beautiful figure components composed of many isosceles triangles?

6. Construct an isosceles triangle with its symmetry axis inside a circle with r = 2cm. Use the distance from the middle point of the circle to the sides as your measure and carry it over many times to the symmetry axis. Connect the component points with the end points of the associated triangle sides. You get a star with many isosceles triangles.

**The Right Angle Triangle**

In form drawing and freehand geometry we also experienced the right angle triangle.[16] We also experienced a special type of triangle that we created with the rope when we produced right-angled triangles. Now as a constructional exercise the class can be asked to do the following:

Exercise: Draw a circle and mark on it the angles of a dodecagon. In order to construct these points draw a horizontal diameter and construct the perpendiculars from it. Carry over from all four end points the circle radius on both sides. We think of the horizontal diameter as the fixed diagonal of all the rectangles, whose second diagonal now moves by virtue of each opposing angle of the dodecagon. Follow the changing form of the rectangle. When does it become a square? Think of the right angle triangles as "half rectangles" above and below the horizontal diameter. Can you see other right angle triangles in the figure? (See Fig. 24.)

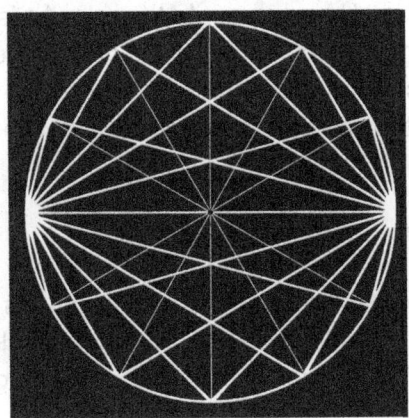

*Fig. 24: Rectangles and Right Angle Triangles in a Circle*

Since we will go into more detail about the right angle triangle later, for now, going over one possible transformation of form should be enough.

**The Scalene Triangle**

If we avoid making any specifications for the angles or sides of a triangle, the triangle can take on completely new forms. Ask the children to draw on the blackboard strange, never-before-seen triangles. Then take a length of rope and let the children hold it at three different points, the circumference area of the triangles remains the same, but they can take on a richer variety of forms. After trying this out, there is a desire for ordering, categorizing terms.

Besides the differences in triangle forms we have already seen, one also divides them according to their largest angle. A triangle is called an:

*Acute Triangle:* When all three angles are less than 90°

*Right Triangle:* When the largest angle is 90°

*Obtuse Triangle:* When the largest angle is more than 90°

Naturally, a right angle triangle can never be acute or obtuse. An equilateral triangle is always an acute triangle. An isosceles triangle can be acute, right, or obtuse. In view of the great variability of triangle forms, it is surprising to find that there are rules that apply to all of them. In the following we will look at the most important rules.

*Exercises Using the Scalene Triangle*

The following exercises are also preparations for perspective drawing in the sixth grade. They show how triangles can be seen differently from different perspectives, reproducing themselves in shadows, as the case may be.

1. Cut a regular triangle (not isosceles or right angle) out of cardboard. Examine its shadow forms. Could they be isosceles or right angle triangles?

2. Take two differently formed triangles. Put the larger one in front of you and hold the smaller[17] one at some distance from your eyes and try to cover it exactly with the larger triangle. Is this always possible? (Yes).

3. A gable on an old house has the approximate form of an isosceles triangle. As you walk by the house, what triangle form can you see between the house and the roof? (If one wishes to pursue this seriously, one can give the children a glass sheet, hold it perpendicular to the line of sight, and outline the contours as they are seen with a felt-tip pen. If one stands to the side of the house, a surprising triangle form appears. Admittedly, the gable form is not gauged according to this perspective view, but rather according to its actual form.)

4. Much of what can be learned from the preceding exercises plays a big role in architecture. For example, an architect must know: How would roof lines look as seen from different distances and directions? Which form has a single surface area when one walks around it? How are different structures concealed when one walks by?

*Fig. 25: Self-Concealing Structures from the fountain of the monastery in Maulbronn, Germany*

## The Sum of the Interior Angles in a Triangle

In preparation for the following one can begin with an isosceles or also an equilateral triangle.

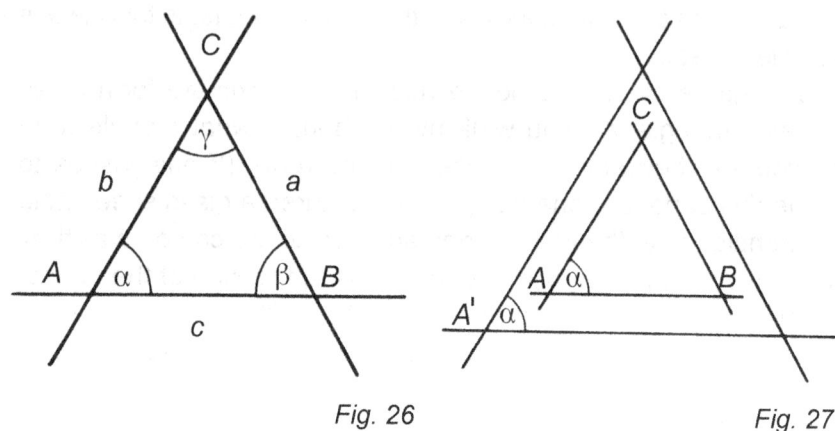

Fig. 26                                                    Fig. 27

After drawing the figure on the blackboard, the children are asked to imagine the figure if it were to move.[18] The straight lines *a* and *b* should rotate around the fixed points A and B; line *c* remains fixed. The children should describe point C's movement if:

- α and β become larger
- α becomes larger and β becomes smaller
- α and β become smaller
- α becomes smaller and β becomes larger

If the children can do a good job of describing the movement of point C, then one may ask a somewhat catchier question: What must one do so that all three of the angles become larger?

Normally, there are children in the sixth grade who then suggest increasing the size of the entire triangle. This provides an opportunity to go over aspects of angles once again: An angle is formed by the *rotation* of a straight line on a point. The number of degrees is a measurement of the rotation. When two triangles have the same form but different sizes, the rotation in A from the direction of AB in the direction of AC remains the same size, just like we would turn the same number of degrees if we were in the corner of the smaller triangle as in the larger. *Angles describe the form, not the length of their legs or the size of the area between them!*

Under no circumstances should one comment negatively about this error that some children are bound to make. As I said, it is an opportunity to return to the discussion about angles and clarify this widely believed (even by adults) misconception about the areas of angles.

In the end, the children will realize that one can never increase or decrease the size of all the angles at the same time, unless one bends the sides of the triangle to the outside or the inside. That gives rise to an interesting point: On a sphere, the sum of the angles of the triangles is more than 180°. This fact plays a big role when airplanes or ships travel around the earth. Once again, the teacher has an opportunity to preview something that will be taught at a later time in the upper grades. One can let the subjects teacher in the upper grades know about this discussion so that he or she can remember to return to it at the proper time and, hopefully, give the children a conscious connection to something they first heard about in the sixth grade.

When the inner flexibility of the triangle has been adequately sampled, then one can bring up again the correct use of the set square as an instrument for measuring angles.[19] Here are three assignments:

1. Draw any triangle on paper and cut it out. Cut the triangle into three parts each containing one corner. Put the corners together and find the 180°.

2. Draw several triangles of your choice in a row, measure the angles and add them up.

3. Draw a triangle with the sides $\overline{AB}$ = 6 cm, α = 40°, β = 50°. What is the size of γ?

The next day we begin by stretching an angle into a straight line and marking a point C on it for the vertex. With two additional lines through this vertex C we can break up this stretched angle in a variety of ways into three angle parts.[20] If we move the first line parallel to itself, then triangles are formed that have the same three angles that we got above when we divided them out of the stretched angle. So, the sum of the angles must remain 180°, the same as in the stretched angle. Since we can produce every triangle form in the same way, then all triangle angles have a sum of 180°.[21]

**Theorem of the Sum of the Angles in a Triangle**
In every triangle the sum of the interior angles is 180°.

$$\alpha + \beta + \gamma = 180°$$

Of course, as proof, we could start with any triangle and, for example, move side c parallel to itself so that it finally goes through C. The angles always stay the same, only the size of the triangle will gradually decrease to zero.

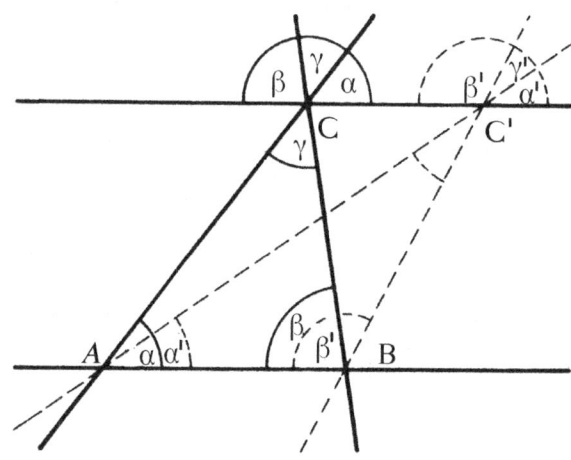

Fig. 28: The Sum of the Angles in a Triangle

It is helpful for the students to modify the figure. For example, move—using different colors—C on the horizontal line through C to the position C'. All angles will change but, the sum $\alpha' + \beta' + \gamma' = 180°$ stays always the same.

## Theorem of the Sum of the Angles in an n-Angle Polygon

This wonderful adherence of any triangle to the theorem also allows understanding and generalization of the sum of angles theorem for polygons. Every convex polygon with n angles (we call it an n-angle) can be divided up into n-2 triangles by using diagonals that begin at one point.[22]

If one starts with one triangle and incrementally adds angles, then the diagonal from a fixed, chosen triangle point to a new point forms a new triangle. Because we started with the triangle, that is n = 3, the number of triangles in the n-angle is n – 2. Since the sum of the angles in every triangle is 180°, the sum of the angles in the n-angle is:

$$S_{IA} = (n - 2) \cdot 180°$$

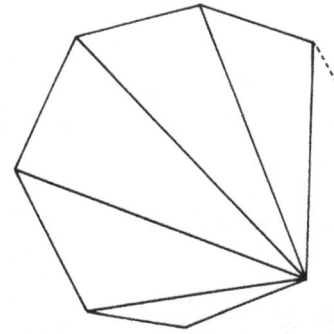 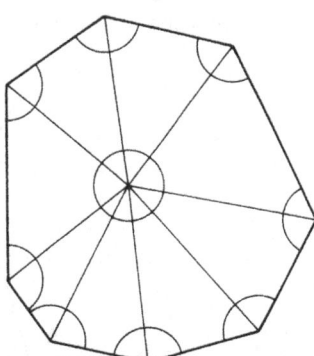

*Figs. 29–30: The Sum of the Interior Angles in a Polygon*

While before this rule was applied only to the regular polygons, now we see it applies to all convex polygons, including the irregular polygons.[23]

When I had shown this to the class, the next day a boy came to me and said he could explain much better why it is (n – 2). He indicated an n-angle, drew an interior point and linked it with all the angles of the n-angle. Now there were n triangles with the total angle sum of n • 180°. However, one must subtract 360° = 2 • 180° from this sum because the full angle of the chosen point in the interior does not belong to the interior angles of the n-angle! How wonderful it is when a young person can discover something like that on his or her own, even if thousands before have made the same discovery!

With this knowledge it becomes immediately clear that an irregular polygon also has the angle sum of $S_{IA} = (n - 2) \cdot 180°$.

## The Exterior Angles of a Triangle

If the theorem of the sum of the angles in a triangle has been adequately discussed, there is something very interesting that can be gleaned from the proof figures: If the parallel line to c is drawn through C, then the angles α and β in C appear in such a way that the exterior angle γ' (it is the completion angle to γ) is the sum of α + β. Therefore, γ' = α + β.

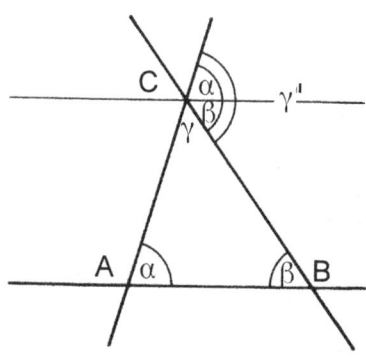

Fig. 31: The Exterior Angles in a Triangle

## Theorem of the Exterior Angles in a Triangle

In a triangle, every exterior angle is the sum of the non-adjacent interior angles.

If we take a special look at an isosceles triangle, we notice the important act that the exterior angle at the apex is twice the size of the base angle: $\gamma' = 2\alpha$. This is because the base angles of an isosceles triangle are the same: $\alpha = \beta$.

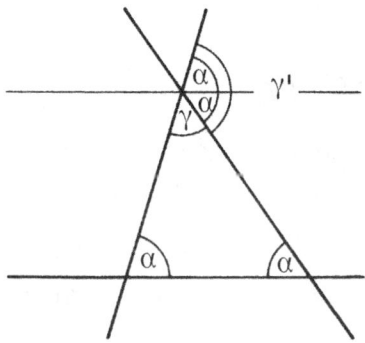

Fig. 32: The Exterior Angles in a Triangle

## Theorem of the Exterior Angles in an Isosceles Triangle

The exterior angle at the apex of an isosceles triangle is the same as the base angles doubled.

$$\gamma' = 2\alpha$$

Note: If the lessons about the triangle have been successful to this point, the children can be given some general, very fundamental information. When we used the rope to walk through a wide variety of triangle forms, it was just a disordered array. Thinking allows us to differentiate and summarize. That is how we learn about the inner richness contained within visible abundance!

From now on the children will become stronger and more independent in their thinking, on their way to adolescence. One can speak with enthusiasm about this approaching time and paint a picture of adolescence for the children so that they see it as something great and beautiful.

## Assignments for the Scalene Triangle

1. In one triangle $\alpha = 41°$, $\beta = 49°$. Calculate $\gamma$. In the same way:

$\alpha = 7°$, $\beta = 118°$, $\gamma = ?$;

$\alpha = 160°$, $\beta = 10°$, $\gamma = ?$;

$\alpha = 11°$, $\gamma = 17°$, $\beta = ?$;

$\beta = 22°$, $\gamma = 48°$, $\alpha = ?$

2. The angles of the bases of isosceles triangles are $\alpha = \beta = 25°$, 30°, 50°, 70°, 85°. Calculate the angles $\gamma$ of the apex.

3. The apex angles of isosceles triangles are given as: $\gamma = 20°$, 40°, 55°, 69°, 100°, 121°, 147°. Calculate the angles of the bases.

4. How big are angles of the bases of a right-angle isosceles triangle?

5. Construct *Figure 33*. Begin with an equilateral $\triangle ABC$. Stretch the side AC along its length to D (double the length of $\overline{AC}$). Connect B and D. Since $\overline{CA} = \overline{CB} = \overline{CD}$; A, B, and D are positioned on a circle around C with the radius $\overline{CA}$.

a) Give the sizes of the angles as $\alpha$, $\beta$, $\gamma$ and $\angle ABD$.

b) Use the figure to construct the perpendicular from the endpoint of a line segment without lengthening it.

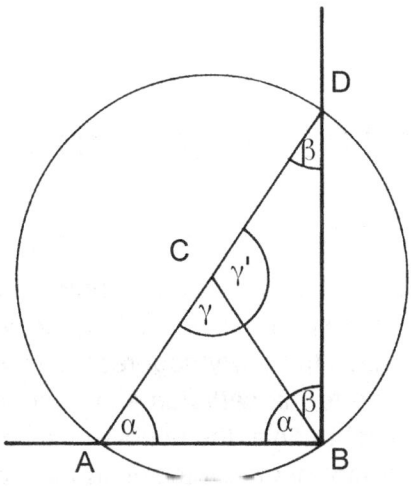

*Fig. 33: For Exercise Number Five*

6. Prove that the bisecting line of the exterior angle at the apex of an isosceles triangle forms a parallel to the line of the base.

7. In an isosceles trapezoid ABCD is $\overline{AB}$ = 10cm, $\overline{CD}$ = 5cm, and the angle of A is $\alpha$ = 70°. Draw the trapezoid and calculate the interior angles. Why must the sum of the interior angles be 360°?

8. What is the sum of the interior angles of any quadrilateral?

9. Construct a regular pentagon[24] inside a circle and draw in the pentagram. Calculate as many angles as you can. Remember that the interior angles of a regular pentagon are 108°.

10. Search for isosceles triangles whose angles on the base are half the size, the same size, or double the size of the apex angle. Can you also find an isosceles triangle whose angle of the base is one-third the size of its apex angle?

# THE RIGHT TRIANGLE AND THE CIRCLE

## The Theorem of Thales

We have also learned about the right-angle isosceles triangle. It is the same as a half of a square. In the fifth grade we discussed it as a special case in relation to the Pythagorean Theorem. At that time we also introduced the terms *cathetus* and *hypotenuse*.[25] The two cathetuses form the right angle, the hypotenuse lies opposite. After we have brought the right-angle isosceles triangle back to mind, we ask: How is one able to change it so that one angle always remains a right angle? Naturally, it can be shifted or rotated. But can we also change its form? For this, we think of the side AB (hypotenuse) as being fixed. How must C be moved so that the angle γ of C remains a right angle?

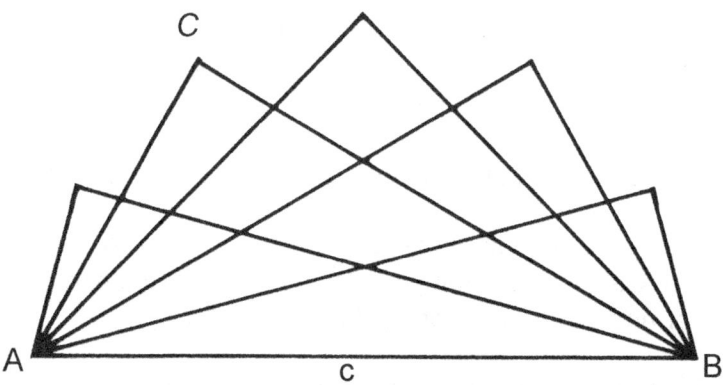

*Fig. 34: Changing a Right-Angle Triangle with a Fixed Hypotenuse*

First, we do an experiment: Place two children in A and B, and ask a third child to walk so that there is always a right angle formed in the line of sight between A and B. One can allow the child to home in on the angle by way of the sides (cathetuses) of a right-angle triangle (set square), A and B, or one can also make an *angle locator.* We could also use a string that is held in tension from A through C to B. C must move in such a way that there is always a right angle formed by the strings. The length of the string does not remain constant! In A or B, the string must be either let out or taken in.

The result is surprising. At first, if one lets the children estimate how C must move so that the right angle is constant, by looking at the rigid figure on the blackboard, there will be every possible suggestion given, but the correct one is seldom fast in coming. What is especially surprising is that C must move along a *bent* line. If the presentation is somewhat largely and carefully carried-out, then a *half-circle* forms above (and below) the line segment $\overline{AB}$.

Is it really true that when one is asked to keep C at a right angle, a circle is formed? Other trials can be undertaken:

*Assignment:* Hammer two nails into a board and let a large enough right-angle triangle go alongside the nails with its cathetuses. The apex of the right angle will show the form of a circle.

As an alternative, one can also mark points A and B on the blackboard, and using a blackboard triangle, move in such a way that both legs of the right angle always go through the points. Follow the apex of the right angle.

Or, one can hold a rod in each hand and rotate them around A and B so that they always intersect at right angles. Follow the intersecting point.

The last example can be transferred to a construction: We draw a sequence of straight lines through A, and from B we drop the perpendicular onto them. This construction is especially nice if we rotate the lines in A at the same angle increments, for example, if we construct a dodecagon or a sixteen-angled figure, or simply make the same angle increments with a protractor (for example, in 15° increments). Those who have drawn the figure exactly can now, at the middle point M of the line segment $\overline{AB}$, draw a circle through all the apexes of the right angles with a compass. This exercise is very suitable as a homework assignment as long as the beginning of the construction has been thoroughly explained.

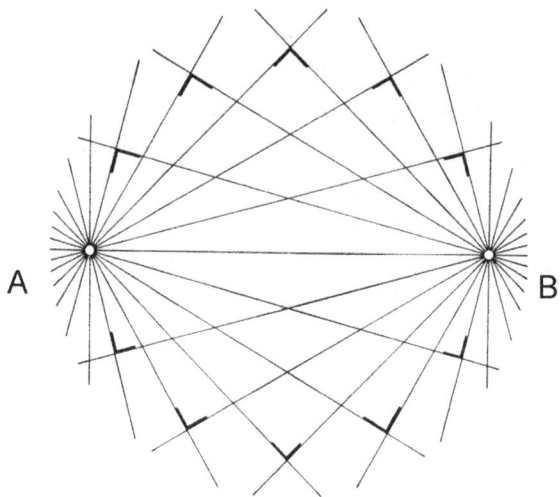

*Fig. 35: Two Right-Angled Pencils of Lines*

The next day, we are not going to try and empirically substantiate the assumption, but rather try and gain the *insight* that a circle is really formed by the right angles.

To this end, again, we draw two points A and B, any straight line b through A, and the perpendicular *a* through B (b should not go through B). The point of intersection of a and b will be called C. We allege that when a and b are rotated, C travels in a circle around the middle point M of the line segment $\overline{AB}$, if the angle of C is always a right angle. We draw the point M and indicate with a dotted line the alleged circle around M through A, B, and C. Who can show why there really is a circle formed?

For a child, it can mean the greatest joy to discover the solution to a problem. There are many possible ways of accomplishing this. We will take a look at two of those ways.

The first possibility requires knowledge of quadrangles and their characteristics as we already studied in the fourth grade in connection with symmetry. The rectangle (including the square) was the only quadrangle whose diagonals were the same length and bisected each other. Once this has been reviewed, or introduced for the first time, then a child can come to the idea of amending the figure into a rectangle through parallels (Fig. 36). The opposing point to C can be named C'. Because a rectangle is formed, the diagonals bisect each other and the other four parts of them are all the same length.

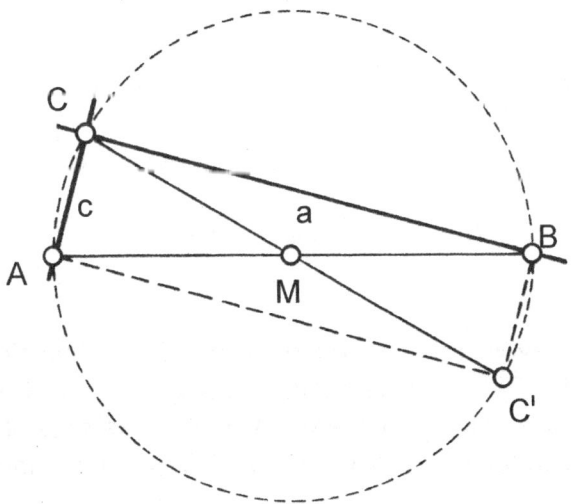

Fig. 36: The Right Angle Triangle Amended to Form a Rectangle

The intersection point of the diagonals must be the middle point M of the line segment $\overline{AB}$. That is, $\overline{MA} = \overline{MC} = \overline{MB} = \overline{MC'}$. If C travels and takes along its right angle, then $\overline{MC}$ must always be the same length, i.e. the same as $\overline{MA}$, which is the same as $\overline{MB}$. C necessarily travels on a circle with the middle point M and the radius $r = \overline{MA}$.

With the second possibility we make use, above all, of the principle of the sum of the angles in a triangle and the characteristics of an isosceles triangle: Since the sum of the angles in a triangle is 180°, and angle $\gamma$ in a right triangle is 90°, then the sum of the other two angles (α+ β) must be 90° also. So, we can divide the right angle into α + β. The dividing lines divide ΔABC into the two isosceles triangles ΔAMC and ΔBCM. Both triangles share the common side $\overline{MC}$. Therefore, $\overline{AM}$ = $\overline{MC}$ = $\overline{MB}$. M is the middle point of the segment $\overline{AB}$ (Fig.37).

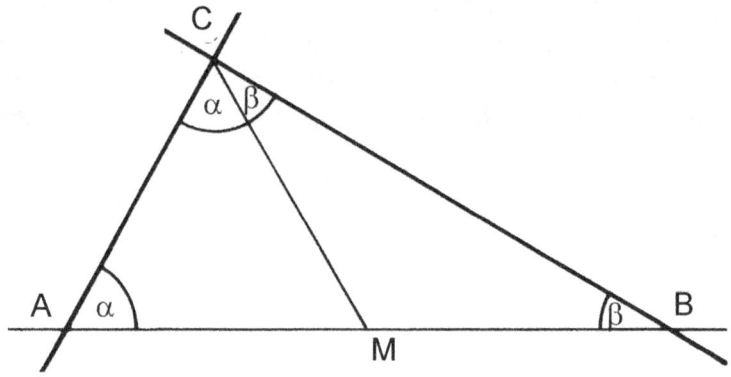

*Fig. 37: The Theorem of Thales: The Right Angle Creates a Circle*

When C travels, both the triangles are always isosceles. Both of its sides, $\overline{AM}$ or $\overline{BM}$, always remain unchanged. C travels on a circle line around M with the radius $\overline{AM}$ = $\overline{MB}$. We take note of the result as being the first part of the well-known *Theorem of Thales*.[26]

**The Theorem of Thales Part I**

If the respective line of two pencils of lines intersects at right angles, then the intersecting point lies on a circle line with the diameter $\overline{AB}$. (The middle point M of the circle is also the middle point of the line segment $\overline{AB}$.[27] If, instead of the pencil lines, we put a triangle ΔABC in the foreground, then we can talk about the Theorem of Thales in the following way:

**The Theorem of Thales I'**

The apexes of all right triangles with the same hypotenuse AB are located on a circle with the diameter AB.

Now, let us look at the theorem when it is *turned around*: Does a triangle, above the diameter of a circle, whose third point sits on the associated circle line, *always* contain a right angle? Even if, at first glance, something else is unlikely, we must admit: Until now we have understood that the right angle creates the circle line in the way that has been illustrated. The other way round, does the circle also, by way of its diameters, always create right angles?

It is of great significance in schooling the children's thinking if they feel justified in not simply believing a conclusion, such as the proof of the Theorem of Thales that we went through, but also wanting to see it proven in the reverse. For example, if I know that all four-year-old children are small, that is a far cry from saying that all small people must be four-year-old children. Or: If I know that all fish live in the water, it does not mean that all animals living in the water are fish. Or: If I say that I always bathe on Saturdays, I have not said that I bathe only on Saturdays.

We draw a circle with a diameter $\overline{AB}$ and the middle point M, and put a point C on the circle. Must the lines a = $\overline{BC}$ and b = $\overline{AC}$ run at right angles to each other? Or, in other words: Must $\triangle ABC$ be right?

After the first proof, the second is already a little easier. If the first possibility has been taken from what is written above, a child' will perhaps suggest that the segment $\overline{CM}$ can be lengthened across and past M to the circle, and there, the opposite point C can be located (see Fig. 38). Because the four points are circle points, they are all the same distance (r) from M. The quadrangle ACBC has diagonals of equal lengths that bisect each other. Only the rectangle has this quality (the square is a special case). So all four angles must be right angles. With that, we have gained Part II of the Theorem of Thales.

For this proof we have used the characteristics of a rectangle for support. However, these characteristics are still completely found in the experience of symmetry and not further substantiated intellectually. For this reason, a second way is given here:

In Figure 38, ΔAMC and ΔMBC are isosceles because M is middle point of the circle and points A, B, and C are placed on the circle. The sum of their exterior angles, α and β of their apexes is 180° and, therefore, the sum of their base angles is 90° (Theorem of the Isosceles Triangle):

$$α' + β' = 180° \text{ and therefore } α + β = 90°$$

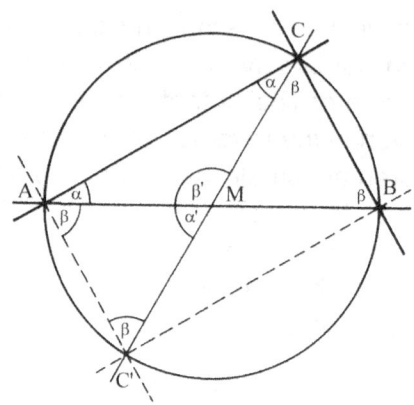

*Fig. 38: The Theorem of Thales: A Circle Creates Right Angles*

Whichever way one chooses, this applies:

**The Theorem of Thales Part II**

If one connects the end points of a circle's diameters with a point on the circle, then both the connecting lines will run at right angles to each other. Or: The angles in a half-circle are right angles.

Let us look at the triangle ΔABC. Then the theorem says:

**The Theorem of Thales Part II'**

A triangle above the diameter of a circle is always right-angled.

As opportunity arises, one can also mention that when we write a *th* in front of Greek names, in the Greek it is written as θ. It is pronounced like the *th* sound in English. Germans pronounce only the *t* sound. Russians speak and write an *f* instead of *th*. They talk about *F*ales and Py*f*agoras.

*Exercises for the Theorem of Thales*

1. Construct a right triangle with the hypotenuse of c = 8 cm and the cathetus of b = 4 cm.

2. Just as in the first exercise, but b should have 1 cm, 2 cm, in succession up to 7cm.

3. Break up the triangle in exercise one into two isosceles triangles.

4. Why must the center perpendicular of the cathetus intersect on the hypotenuse?

5. The diagonal length d = 7 cm, and the side lengths $\overline{AB}$ = 4 cm of a rectangle are known. Construct the rectangle. Use the Theorem of Thales.

6. Draw a horizontal straight line segment and mark two points on it that are 10 cm apart. Erect the perpendiculars of both points. Bisect all the right angles. Bisect all the created angles another two or three times in such a way that all the created angles are the same. Color in the formed quadrilaterals like a chessboard. This exercise, which was introduced in Volume 2, should now be brought into relation with the Theorem of Thales I. It will be simultaneous preparation for the Theorem of Peripheral Angles (Fig. 35).

7. With the help of the Theorem of Thales it is easy to construct a triangle with angles of 30°, 60°, and 90°. They are often used as figure triangles (Fig. 39). Notice that in the given construction, the construction of an equilateral triangle is hidden. It is just as easy to construct a right-angle isosceles triangle with its 45° base angles. The set square has this form, for instance (Fig. 40).

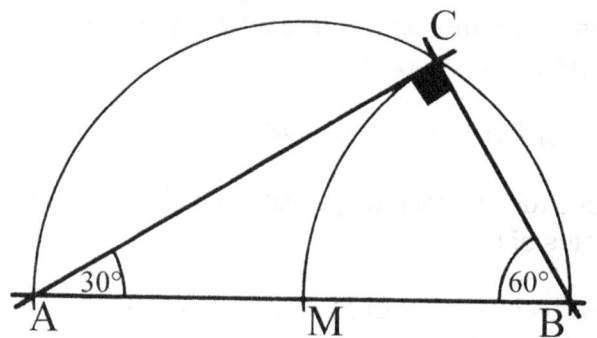

*Fig. 39: Construction of a 30°–60°–90° Triangle*

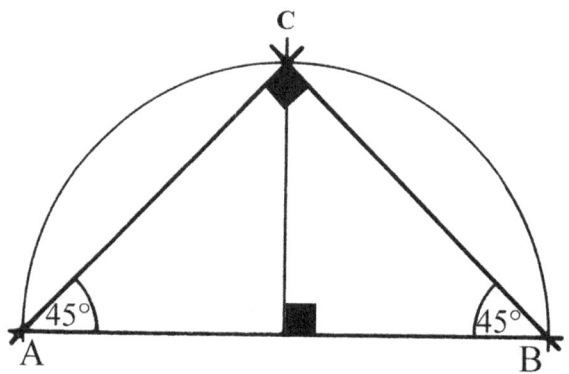

*Fig. 40: Construction of a 45°–45°–90° Triangle*

8. The perpendicular of point P in a straight line g can be erected using the Theorem of Thales as well. We draw a circle around an appropriate point M that is outside of g, through P, and that will intersect g at a second point Q. The central line through Q intersects the circle at a second point R. PR is the perpendicular (Figure 41).

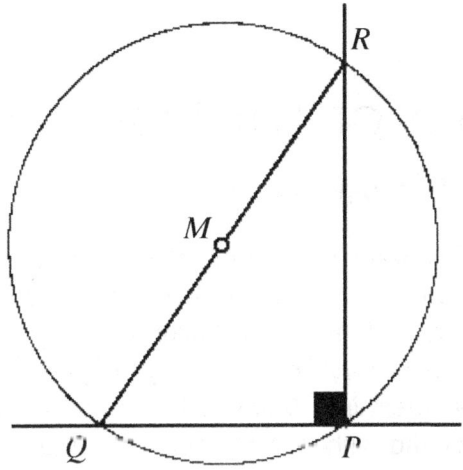

*Fig. 41: A New Construction of the Perpendicular*

9. With the help of the Theorem of Thales, drop the perpendicular from a point P onto a straight line g. Describe the construction.

10. Construct a square with a diagonal length of 8cm. Use the Theorem of Thales.

# The Rules of the Right Triangle

## The Theorem of Pythagoras

Probably the best known theorem of mathematics is the *Theorem of Pythagoras*. Like no other bit of mathematical knowledge, it has accompanied mathematical thinking through more than two thousand years of history and, at every stage, been given a new form and a new meaning. Many cultures in the world know the principles associated with it. No other mathematical theorem has so many proofs, whose special characters reveal much about their sources.[28] The role that the Pythagorean Theorem will play in the future, and what form and significance will be given it by future mathematicians, cannot be overlooked (see Fig. 42).

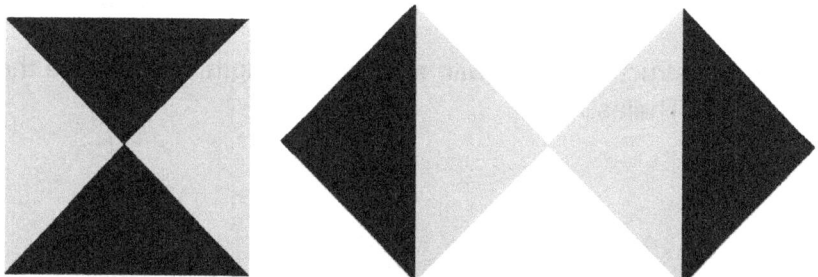

*Fig. 42: A square can be divided into two smaller squares.*

In the fifth grade we were introduced to the Pythagorean Theorem by way of a special form, namely the isosceles right triangle. If it was not yet introduced—we introduce it now. We start with the question: How can we transform a given square into two smaller squares each with the same area? One possible answer is shown in Fig. 42: Cut the given square by the diagonals into four right isosceles triangles and compose out of them two smaller squares. Notice that the area is not changed.

As mentioned in vol. 2 we can relate this figure to the traditional figure of the Pythagorean Theorem by talking about isosceles right triangles and squares in Fig. 42 and looking at them from a different point of view. When we combine the two parts of Fig. 42 we get: In a right isosceles triangle is the total of the cathetuses[29] squares equal to the square of the hypotenuse (Fig. 43).

*Figs. 43 and 44: The Pythagorean Figure*

Now we can ask the question: Can the given large square also be transformed into two different squares leaving the total area unchanged? It is not self evident that the answer can be found using right triangles. But we can suggest it.

Just as we could transform the right triangle using the Thales Theorem, the same can be done with the Pythagorean figure. What would happen to the theorem if the vertex of the right angle moved on the Thales circle? The figure can help to incite this idea of a flexible figure. Therefore we draw above the hypotenuse of the given right and isosceles triangle a semicircle. If the vertex of the triangle moves along the semicircle, we know that the angle will always be a right angle. Of course, the cathetuses (the two legs of the right angle) and their associated squares will change; meanwhile the hypotenuse square remains fixed (Figs. 44 and 45).

The children are now asked to imagine the whole Pythagoras figure in movement. It needs some effort but, it is very beautiful to make after mental imaging a drawing with regular distances on the semicircle as in Fig. 45. In this figure the hypotenuse square is drawn above the base because the wonderful lawfulness in the relation between it and the cathetuses squares becomes more visible. Meanwhile the square of the hypotenuse remains fixed (as said before) and those of the cathetuses change, as can be seen in Fig. 45. If one of them is shrinking, then the associated will be growing. Is the loss of the one balanced by the gaining of the other?

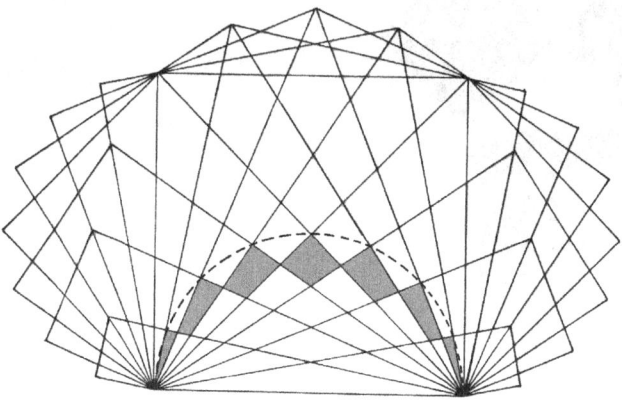

*Fig. 45: The Pythagorean Theorem*

One special case—beside the isosceles case—is if the vertex meets one endpoint of the hypotenuse. In this case the one cathetus will be equal to the hypotenuse and the other will be null. So the total of the cathetuses' squares will be equal to the hypotenuse square. Will this hold true for all positions on the semicircle?

The following proof should be done with and by the children as vividly as possible with pieces of colored cutout cardboard. The starting point is the Pythagorean figure (Fig. 44). In order to find out whether the areas of the cathetuses squares fit into the hypotenuse square, we once again turn the latter above the hypotenuse and cut off all parts of the cathetuses squares that still can be seen (Figs. 46 and 47). Now, we will try to insert all parts into the hypotenuse square.

My suggestion is to let the children work individually or in groups with the different parts for a longer time. Don't push them forward!

This is a wonderful exercise with different shaped areas and possible results are anticipated by inner pictures and the process of playing with different variations. Exact inner picturing is trained with this exercise.

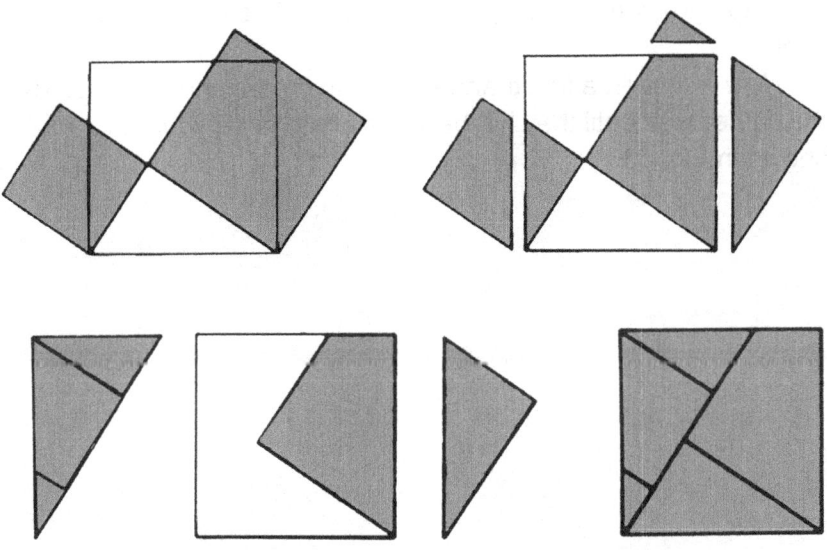

*Figs. 46–49: First Proof of the Pythagorean Theorem*

Next day, when the children have had the opportunity to find an answer (Figs. 48 and 49), we can work more consciously on the question: *Why* is the answer correct? Remark: If this part seems to be still too difficult don't hesitate to skip and come back later to it.

I will show here how a proof can be done without being too abstract for a 6th (or 7th) grade. Our starting point is, again, the Pythagorean figure that we have carefully drawn on the blackboard. We label it $c^2$ (Fig. 50).[30] Now we add a proper right triangle and label it as shown in Fig. 51.

By adding this triangle the symmetry of the figure is disturbed. We balance this disturbance by adding three more triangles equal to the first one as shown in Fig. 52. It seems that we got a new square. Is that really true? That is easy to understand: Just look at a corner of $c^2$. The three angles which appear there are the three angles α, β, γ of the added triangles. Because the total of the angles in a triangle is 180°, the line labeled a + b is straight. Otherwise this line would be bent and we would have an octagon instead of a quadrangle. Now we make clear that the four sides have the same length a + b. So the quadrangle is a square. By the addition of the four triangles we got a symmetric figure that we will meet again later.

Now we insert a fifth triangle equal to the other ones and extend the cathetuses until they hit the sides of the large square (a + b)² as shown in Fig. 53.

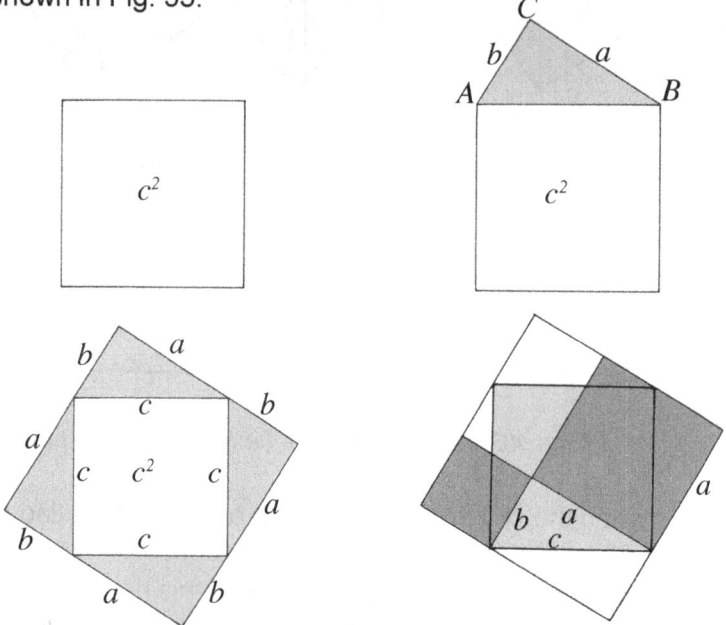

*Figs. 50–53: First Proof of the Pythagorean Theorem I*

Exercise: Label Fig. 53 carefully and completely. Where else can a, b, c be found? Insert also α, β, γ.

Our question of which we should never lose sight is: *Why can the different parts of the cathetuses squares that we have in Fig. 47, and similarly in Fig. 53, be inserted into the hypotenuse square in such a way that it is absolutely covered and nothing of the cathetuses squares is left over?*

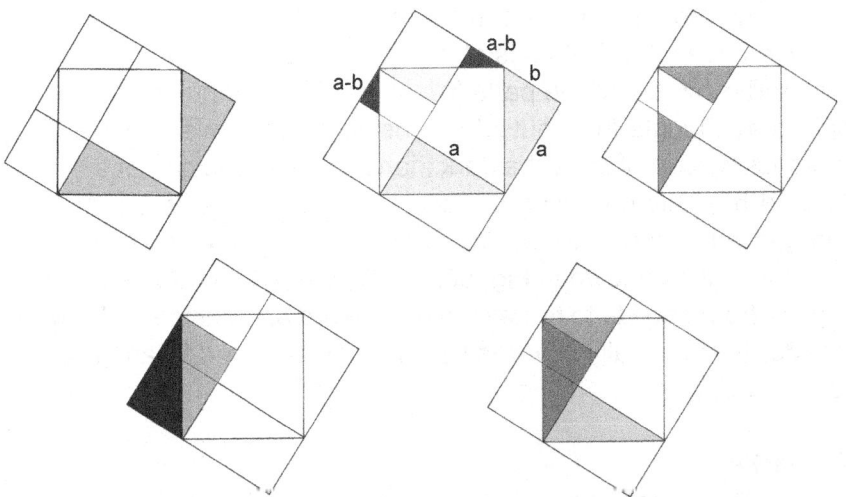

*Figs. 54–58: Second Proof of the Pythagorean Theorem*

We will explain this step by step using a large drawing on the blackboard.

1. The light gray triangles (please use colors instead of different grays) in Fig. 54 are equal because we added only equal triangles. Therefore we can put the outer triangle inside $c^2$.
2. The two dark triangles in Fig. 55 are equal because they have equal angles and at least one equal length of one cathetus (a-b).[31] We come back to this identity.
3. The middle grey shaded triangles in Fig. 56 are equal because they arise from the same partition of $b^2$. This is true because if we drop a perpendicular from the upper left point of $c^2$ on the extension of b, we get $b^2$ with the same partition.[32]
4. Now, we still have to make clear that the two shaded triangles in Fig. 57 are equal. This holds true because they are halves of a rectangle. So, the darker shaded triangle outside of $c^2$ can be inserted inside $c^2$.

Now, we combine all individual steps and get Fig. 58: All parts of the cathetuses squares cover in the shown way the hypotenuse square and nothing is left over.

Let us now repeat the different steps: We start with Figs. 46 to 49 where we cut off all parts of the cathetuses squares that are not covered by $c^2$. Do all the parts fit into $c^2$? Then we put in Fig. 54 the light gray triangle from outside to inside $c^2$. Then we complete the out-sticking part of $b^2$ by the dark triangle which was part of $a^2$ (Figs. 55 and 57). This produces a triangle that is equal to the light gray triangle in Fig. 57. Then we move that part of $b^2$ that is inside $c^2$ up and rotate it as shown in Fig. 56. Finally we put the dark triangle in Fig. 57 from outside to remaining uncovered area inside $c^2$. Now we are done: $c^2$ is totally covered by the parts of $a^2$ and $b^2$ and nothing is left over.

**Remarks:**
1. This proof has, of course, some formal weaknesses. But, after reviewing the following chapter on congruent triangles, it is not difficult to be more precise. In general, an axiomatic system of geometry is not yet appropriate; that can be brought in the high school. We still are here on a propaedeutic but nevertheless important level. My intention is to offer a really beautiful geometric content in such a way that the need for causality of the 12-year-old students can be met. The arguments presented here can satisfy their thinking without a too abstract or formal level that might result in an aversion to geometry.
2. Written text cannot substitute for oral presentation with the movements of our hands to explain the shift of areas, the colors, and, above all, the vivid discussions with the students, their questions, doubts and suggestions.
3. Reading the text for the first time may look very complicated —maybe even too complicated—for sixth graders. It is not. Just the focus on an overview of the individual different steps and keeping the target in sight can be very educational at this developmental stage. Not least because of educational relevance Rudolf Steiner suggested this proof in his lectures for the future Waldorf teachers.[33]
4. For the sake of clarity let me add: The experimental experience with cardboard as shown in Figs. 46 to 49 is

just the beginning of a proof. The exercise does not yet give the inner necessity of the truth that we can reach only by an inner process called *thinking*.
5. The proof and even how to fit the parts of the cathetuses squares into the hypotenuse square will often be forgotten. This is quite natural, and it gives us the chance to convince oneself again and again of the truth and maybe to find other ways for the argumentation or other breakups of the squares.

**Supplementary Proof (the Indian Proof)**

Because there are many ways to get insight into a mathematical law we add a second very beautiful proof, the so-called Indian proof.[34] Thereto we put the figure of Pythagoras with colored cardboard in two different ways on a table or fix it at the blackboard—once the right triangle with only the cathetuses squares and once with the hypotenuse square (Figs. 59 and 60). Beside the first two triangles we have six more in store. Now we ask: Who can add the triangles in my hand to the given figures in such a way that they become more beautiful?

While not very precise mathematically, this question has, in my experience, always produced a solution, for example, as shown in Figs. 59 to 62. In both figures the same-sized squares come about. This can be quickly understood when one looks at the total of the angles that are found in the meeting points of various areas, and when one compares the sides with the length a+b as we already did in the beginning of the first proof.

From here, the truth of the Pythagorean Theorem can be immediately recognized: If we take away from Figure 61 (left bottom) triangle four times, then the area of the cathetuses squares remains. If we do the same thing in Fig. 62 (bottom right), then the area of the hypotenuse squared remains. Taking away equal from equal produces the same remainder.

With these proofs, we can formulate the famous Theorem of Pythagoras anew:[34]

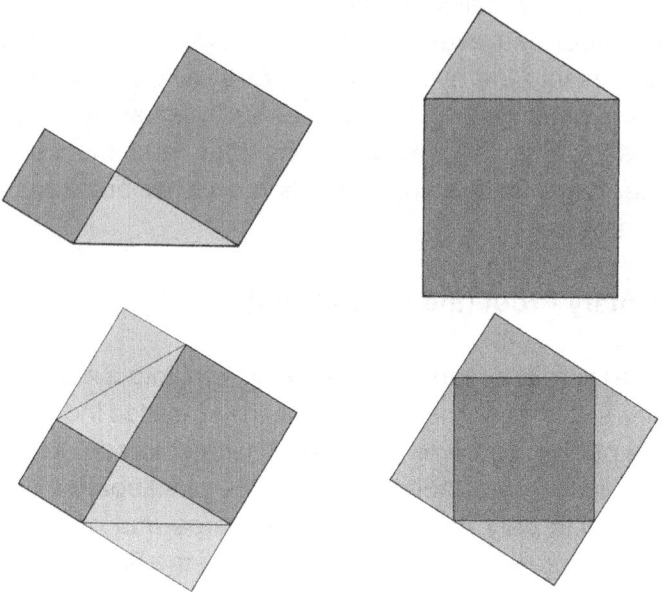

*Figs. 59–62: An auxiliary proof of the Pythagorean Theorem*

**Pythagorean Theorem**
In every right triangle the hypotenuse squared is equal to the sum of the cathetuses squared.

Tip on method: In *The Study of Man* Steiner points out the significance of the imagination in this age group:

> Thus all the teaching, even what is given in geometry and arithmetic, must consistently appeal to the imagination. We appeal to the imagination if, in dealing with plane surfaces, for instance, we endeavor (as we have been doing in our practical course) to make them not only comprehensible to the intellect, but to make them so thoroughly comprehensible that a child needs to use his imagination even in arithmetic and geometry. That is why I said yesterday that I wondered that nobody had thought of explaining the theorem of Pythagoras

in the following way. The teacher could say: "Suppose we have three children; the first has just so much powder to blow that he can make it cover the first square; the second so much that it will cover the second square; the third so much that it will just cover the little square." We shall be helping the child's imagination when we show him that the powder needed to cover the largest square is the same in quantity as that needed to cover the other two squares. Through this the child will bring his power of comprehension on the powder blown on the squares, perhaps not with mathematical accuracy, but in a form filled with imagination. He will follow the surfaces with his imagination. He will grasp the theorem of Pythagoras by means of the flying and settling powder, that would have to be blown moreover into square shapes (a thing impossible in reality of course, but calling out the exertion of imagination). He will grasp the theorem with his imagination.[35]

Through the imagination a child connects much more intensively with the study material. Interestingly, the imagination is especially stimulated when illustrated situations are not taken from everyday life, but rather are paradoxical or factually impossible.

Closer to reality, one can think of a garden area that needs to be dug up. Whoever has dug up the area of the hypotenuse squared will have done just as much as the one who dug up the area of the sides squared. Similarly, pieces of pizza will satiate equally well when they fulfill the conditions of the Pythagorean Theorem.

With the help of the Pythagorean Theorem, a square can be divided into two squares whose sum equals the original the original square. Example: A square has the side length $c = 10$cm. Divide it into two squares in which the side length of one is $a = 6$cm. Measure the side length $b$ of the other square.

Solution: First, we draw the given line segment c = 10 cm, construct its middle point M, and draw a semicircle around M through the end points A and B. Then, we put the line segment a = 6 cm in a compass and draw a circular arc around B that intersects the half-circle. The intersection point C is the third point of the triangle, whose sides give the side lengths of the divided square. If you worked correctly you will find that b = 8 cm (Fig. 63).

2. Example: Two squares with the side lengths a = 7 cm and b = 4 cm are given. Construct a square with the total area of both given squares. Measure the side length of that square.

Solution: We start with one of the given sides, e.g. a = 7 cm and construct the perpendicular segment b = 4 cm in one endpoint of a. Connect both free endpoints of a and b by c. The desired square is $c^2$ and c is a little bit more than 8 cm (Fig 64).

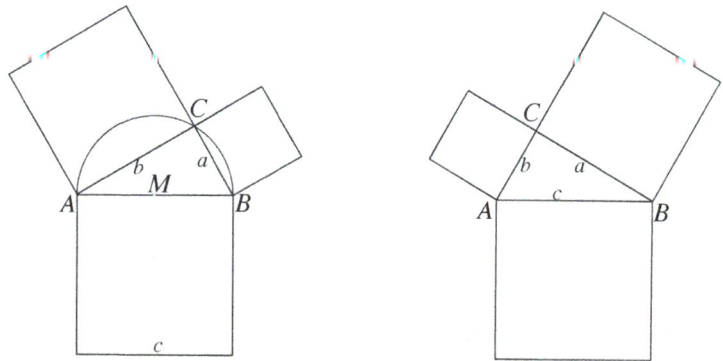

Figs. 63 and 64: Dividing and Joining Squares

## The Reverse of the Pythagorean Theorem

In order to lead backwards into the theorem, we ask: Does the Pythagorean Theorem also apply to a triangle that is *not* right-angled? I believe that a strict, indirect proof would go right over the heads of most sixth graders. However, by using the Pythagorean figure, one can make clear that the above is not possible. Namely, if a *non*-right triangle △ABC had one side squared equal to the sum of the two other sides squared, one could, for instance, drop the perpendicular from A onto where a or its prolong it hits C'. In △AC'C is $\overline{AC'} < \overline{AC}$. Because the Pythagorean Theorem applies to △ABC, it cannot equally apply to △ABC'. (The teacher must determine what of the above can be brought to his or her specific class.)

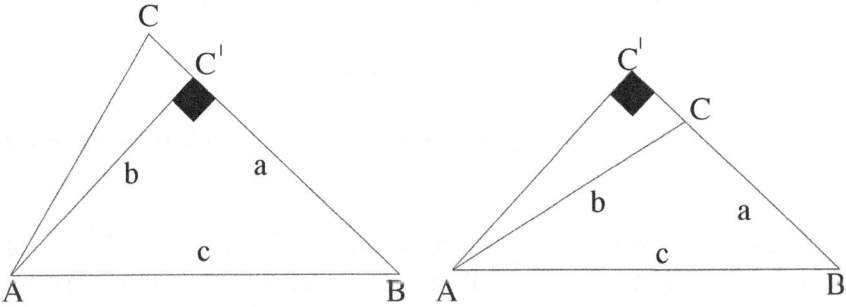

*Figs. 65 and 66: Reversing the Pythagorean Theorem*

In every case, the **Reverse of the Pythagorean Theorem** applies: In a triangle, if the sum of the squares of its two sides equals the square of the third side, then it is a right triangle. In the upper grades it will be shown what relationships exist between the sides of a triangle that is *not* right-angled.[36]

What before was a theorem treated in a purely geometric way is now also to be connected to *measured values.* To do this, one recalls, or introduces for the first time, that the area of a square can be easily calculated if the side lengths are measured: If the segment c is measured by a unit of length, a cm, for instance, then the measure of the square area is a x a units of area, whereby one unit of area is considered as a small square with the unit of length as the side lengths. In order to express this relationship between units of length and area, one writes the units of length like this, for example: $cm^2$, or $m^2$, or $ft^2$, etc., and reads it as square centimeters, square meters, or square feet, etc. The "2" expresses that the size of the area is expanded in *two* dimensions.[37]

Although it will be discussed in seventh grade algebra in more detail, one can also write the product $a \times a$ as $a^2$. For example: If $c = 5$ cm, then $c^2 = 5^2$ cm$^2$ = 25 cm$^2$.

If we have a right triangle with the cathetuses a and b, and the hypotenuse c, then according to the Pythagorean Theorem:

$$a^2 + b^2 = c^2$$

This applies to the equality of the areas as well as their measured value.

A triple of natural numbers a, b, and c, fulfilling $a^2 + b^2 = c^2$, is called a *Pythagorean Triple*. If one constructs such a triangle with the sides a, b, and c, then it is a right triangle. Examples are 3, 4, 5 and 5, 12, 13 because $3^2 + 4^2 = 9 + 16 = 25 = 5^2$ and $5^2 + 12^2 = 25 + 144 = 169 = 13^2$.

This knowledge was used by the Ancient Egyptians in order to achieve exact right angles for the foundations of the pyramids in the desert sands. In fact, *geo-metry* (= earth measurement) played a large role in Egypt because the Nile River, with its yearly floods, not only left fertile silt on the fields, but also often washed away the measured boundary lines. Herodotus wrote: "King Sesostris was said to have divided the land among all the residents and given each an equal, square piece of land. The yearly rent he collected was the substance of his income. If the river current took something away from someone's acreage, then the owner would go to the king to give notice of it. The king would send people out to measure the loss so that the owner would have to pay rent only for the remaining acreage. It seems to me that geometry was discovered in this way and later brought to Greece."[38]

*Fig. 67: The Great Pyramids at Giza*

Ernst Bindel writes: "The significance of geometry went much further than specific instances of practical land measurements, namely, when it came to constructing their temples. In this case, the right angle came especially into use. The correct orientation of the temple building required a painstaking adherence to the right angle. At the beginning of construction, it was celebrated in a ritual in the form of staking out the foundation. Before the right angle was incorporated into the ground, pegs were driven whose connecting lines marked the baseline of the structure. This part of laying the foundation stones is reported in writings in the Horus Temple at Edfu in Upper Egypt. The Pharaoh himself performed this important act, it is said, together with the Goddess Sekhmet, on the basis of an observation of the sky which gave him the intended directional orientation: "I hold the peg and the sledge, I hold the rope together with the Goddess Sekhmet. My gaze follows the path of the stars. When my eye has come to the constellation of the Big Bear, and the numbers on the dial are showing the proper segment of time, then I will lay out the corners of your holy temple." According to Cantor, these are the words used by the Pharaoh written about in the inscriptions to carry out the above-described task. In one hand he held a club with which he drove a long stake into the ground, and, opposite him, the Goddess Sekhmet, mistress of laying foundations, did the same (Fig. 68).

*Fig. 68: The Pharaoh and the Goddess Sekhmet span a rope.*

The second act, carried out on the orientation of the baseline, was the laying of the right angle by the "harpedonaptae," or "surveyors." These Ancient Egyptian surveyors, who belonged to a special trade guild, would hold a measuring rope that showed twelve equal lengths by way of very artistic and exact knotting or stamped markings. So, the measuring rope consisted of a 3-segment length, a 4-segment length, and a 5-segment length consecutively. The 4-segment length was put on the baseline that had been determined. The 3 and 5-segment lengths were bent up from the ends of the 4-segment length, and, by tightening the rope, the two free ends were joined together. A right angle was formed between the 4-segment length and the 3-segment length. The rope, spanned and bent in this way, inscribed the form of a right triangle into the ground, a form which was held to be sacred (Figs. 69–70).[39] This applies to the triangle with the given sides:

$$3^2 + 4^2 = 5^2$$

We can complete the construction with the help of a rope.

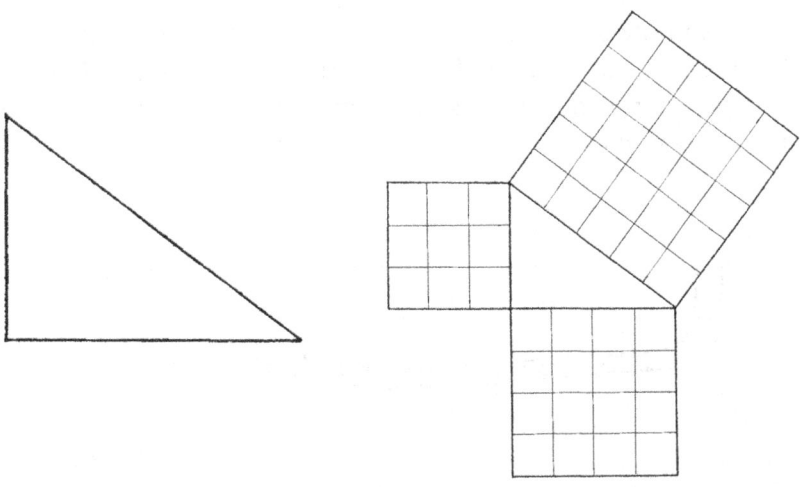

*Figs. 69 and 70: The Sacred Triangle of the Egyptians*

Of course, the construction of the right angle in the desert sand was not the end of the task of having the foundation square exactly at a horizontal level. This was of the utmost importance to the stability of the pyramid and the proper fitting together of the blocks. One can illustrate how this could have been done:

In the house-building block of the third grade the children become familiar with the tool known as a level. Using a very large "level," one can create the exact horizontal and level form for the foundation square. With the help of a knotted string and the spacing measurements, one determines the position of the corners and stakes them out in the sand. Then one would dig ditches between the corners and line them with Nile mud or clay, and let water flow into them from the irrigation canals dug from the Nile. Or, perhaps, many men were put to work passing skins filled with water to fill the ditches. At the point of the highest water level on the ground stakes, a notch was made to mark the highest water level because it would sink again rapidly. Finding the exact measurement was long, hard work. However, much that was to be done later depended on the exactness of the work in the beginning. Had the Egyptians possessed hoses they could have also used a hose level. A hose is filled with water. If one holds both ends at the same height then no water will run out. In this way, one can, over larger spacing, determine when two points are at the same height.[40]

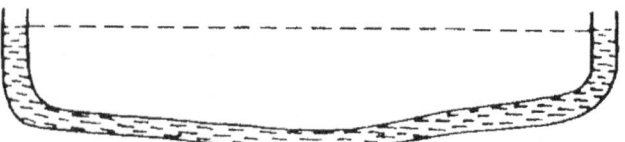

Fig. 71: The Hose Level

*Assignments Using the Pythagorean Theorem*

1. Divide a half-circle into six equal segments. Draw all the associated right triangles above the diameter, and construct the side squares for each triangle.

2. A square with side lengths of 10 cm is given. Deconstruct it into two squares, one of which has the side lengths of 3cm. Measure the side lengths of the other square.

3. Show by calculation that triangles with the side lengths of 6, 8, 10, and 11, 60, 61, are right-angled.

4. Create a table with the square numbers of $1^2$ to $25^2$. Besides 3, 4, and 5, can you find other Pythagorean Triples in this group?

5. Two squares with side lengths of 4cm and 5cm are given. Construct a square with the areas of both of the single squares.

6. Show through consideration of the angle that a triangle which can be taken apart into two triangles that have the same angles as the original triangle, must be right-angled.

Solution: If we want to deconstruct a triangle △ABC with the angles (alpha α, beta β, gamma γ) into two triangles, the dividing line has to pass exactly though one angle, e.g. C. Let us say this line hits c in point D. The two new triangles are △ADC and △BCD. They are required to have the same angles (α), (β), and (γ). If we check where (γ) can be only D is the point where it can exist. Because the angles in D have to be equal (γ) has to be 90°. This is possible only if the original triangle is right-angled (Fig. 72).

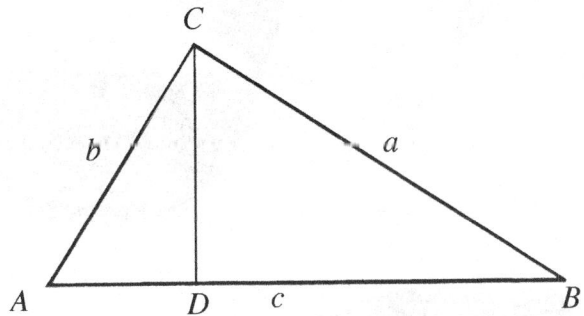

*Fig. 72: For Problem 6*

**Addendum to the Pythagorean Theorem**

If the sum of the squares of the sides equals the hypotenuse squared, then half, or another fraction of the square, must also have corresponding identical areas. An example of how the Pythagorean Theorem can be expressed also with other forms of area is given in Figs. 73–75. In the seventh grade it will be shown that equality of all corresponding areas applies only when they have the same form.[40]

*Figs.73–75: The Pythagorean Theorem also applies to elephants and other forms.*

# The Principles of Congruence
## and Basic Assignments Using the Triangle

We have learned about the regular triangle and its great variability. If one wishes to ascertain the form and size of a certain triangle (not its position), one need not provide all three of the sides and angles. For instance, if the three sides are given, then one sees immediately the form and size of the triangle. If, on the other hand, two angles are known, then the third angle, according to the principle, is determined by the sum of the angles in the triangle, but the *size* of the triangle is still unknown.

At this point, one should consider, with the class, a certain triangle in which the form and size can be determined. We assemble the four, or possibly five, cases with their prerequisites. In order to understand this without a proof, one thinks of a second triangle, with the same length segments as the first, laid on top of the first triangle in a corresponding way. By doing this one starts with an appropriate number of equal segments, then the rest of the segments also have the same position. We also look at two triangles as congruent when one has to be turned around, i.e. mirrored, in order to cover it.

**The Principles of Congruence**

Triangles are congruent (equal) when they correspond:
1. In one side and both adjacent angles
   (The condition for the given sizes is that the sum of the given angles is less than 180°.)
2. In two sides and the enclosed angle
   (The condition is that the given angle is less than 180°.)
3. In two sides and the angle opposite the larger side.
   (The condition is, again, that the angle is less than 180°; if the angle opposite the smaller side is given, there are *two* solutions.)

4. In all three sides
(The condition is that the sum of both of the smaller sides is more than the third side.)

## Basic Assignments Using the Triangle

The basic assignments using the triangle come from these four cases, which the children should certainly master. One speaks of basic assignments because only sides and angles are given. Later, there will also be exercises with more difficult determined factors given. For every type of assignment an example is given.

With every assignment, the individual sizes can be determined by *figures* (geometric) or by *given measurements.* If the segments of a triangle are determined by given measurements, then a ruler and a set square should be used.

1. a) In the triangle △ABC, side c and both adjacent angles α and β, are given. Construct the triangle.

*Fig. 76: Example for Assignment 1a.*

Solution: First we draw line or side c in the desired position and transfer angles α and β to the end points A and B on the same side of c.[41] The free sides from α and β intersect at the third point C.

1. b) In the triangle △ABC the side c, the adjacent angle α, as well as the opposite angle γ is given. Construct the triangle.

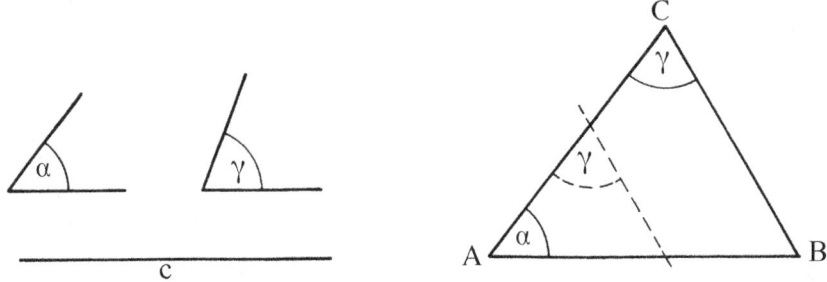

*Fig. 77: Example for Assignment 1*

Solution: First we make α in A; then we put γ, located at any place on the free side of α, at the side where side c is located, and shift the second side parallel in such a way that it goes through B.

(Additional: Who can find another solution? β = 180° - α - γ)

2. In the triangle △ABC, the sides a and b, as well as the enclosed angle γ are given. Construct the triangle.

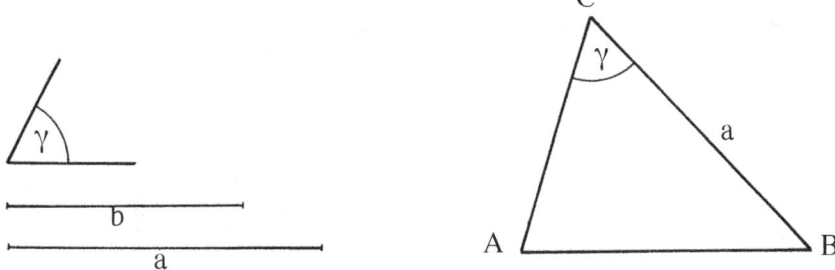

*Fig. 78: Example for Assignment 2*

Solution: We put one side in the desired position, a, for example. We put angle γ at an end point C, and give the free side the length b. Side c connects the still unconnected end points of sides a and b.

3. a) Two sides and the angle opposite the larger side of a triangle are given. Construct the triangle.

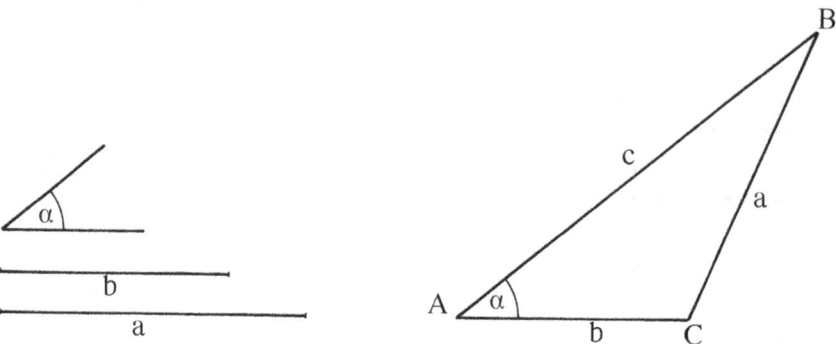

Fig. 79: Example for Assignment 3a

Solution: First we draw the smaller of the two given sides, b, for example. One of its end points we designate as A and put the given angle α there. We put in the second given side by Figure a circular arc around the other end point C of b, with the radius a it intersects the free side of α in B. $\overline{AB}$ is the third side c.

3. b) Two sides, and the angle opposite the smaller side of a triangle are given. Construct the triangle.

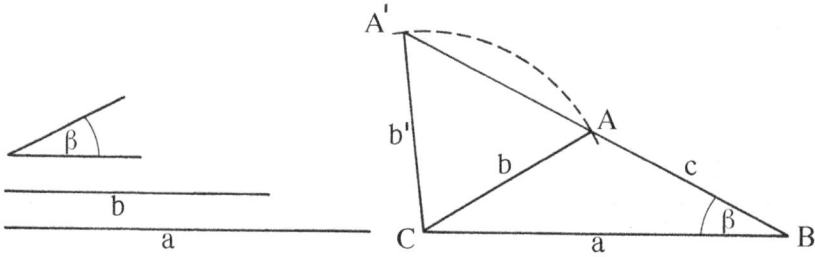

Fig. 80: Example for Assignment 3b

Solution: This assignment has *two* solutions; in this case the triangle is not clearly determined. Both solutions can be accordingly found for 3a, but then one would start with the larger of the two sides.

4. In the triangle ΔABC three sides are given. Construct the triangle.

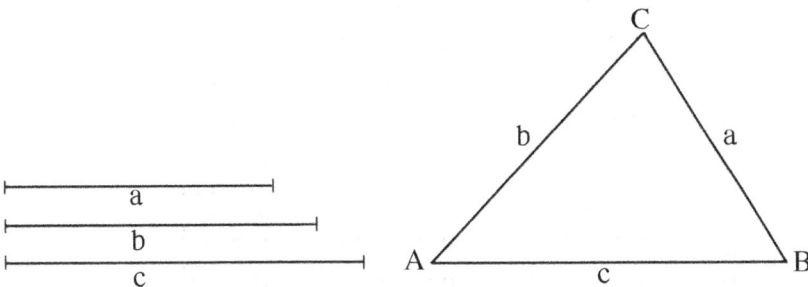

*Fig. 81: Example for Assignment 4*

Solution: First we draw one of the three sides and put a circle line around one end point, with one of the other sides as the radius and another circle line around the other endpoint with the third side as the radius. The intersection points of the two circle lines (radii) give both possible solutions, two mirror-symmetrical, congruent triangles.

## Additional Assignments Using the Triangle

1. Construct the following triangles:

   | | | | |
   |---|---|---|---|
   | a) | a = 8 cm | β = 42° | γ = 80° |
   | b) | b = 7 cm | α = 20° | γ = 140° |
   | c) | c = 9 cm | α = 16° | β = 27° |
   | d) | a = 7.2 cm | α = 39° | β = 60° |
   | e) | b = 6.6 cm | α = 55° | β = 95° |
   | f) | c = 9.1 cm | β = 100° | γ = 30° |
   | g) | a = 4.5 cm | b = 10 cm | γ = 40° |
   | h) | b = 10 cm | c = 8 cm | α = 23° |
   | i) | a = 5 cm | c = 4.5 cm | β = 25° |
   | j) | a = 6 cm | b = 4 cm | α = 85° |
   | k) | b = 9.8 cm | c = 4.8 cm | β = 122° |
   | l) | a = 9 cm | c = 4 cm | α = 140° |
   | m) | a = 7 cm | b = 4.8 cm | β = 25° |
   | n) | b = 5.6 cm | c = 4 cm | γ = 37° |
   | o) | b = 9 cm | c = 8 cm | γ = 64° |
   | p) | a = 6 cm | b = 4 cm | c = 3 cm |
   | q) | a = 6 cm | b = 4 cm | c = 2 cm |
   | r) | a = 5 cm | b = 5 cm | c = 3 cm |

2. Points A, B, and P are given. The segments $\overline{PA}$ and $\overline{PB}$ can be measured, but the segment $\overline{AB}$ cannot be measured. For instance, the land is impassable because of a swamp, a lake, or a wheat field before harvest. However, $\overline{AB}$ can be determined with the help of Fig. 82. As an example, choose $\overline{PA}$ = 130 meters, $\overline{PB}$ = 170 meters, and γ = 95°. Construct △ABP in the proper scale and measure $\overline{AB}$. Recalculate it in the original scale.

3. How can one determine the width of the river $\overline{AB}$ from the riverbank (Fig. 83)?

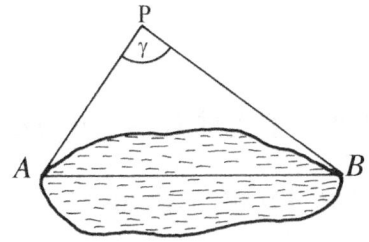

*Fig. 82: For Assignment 2*

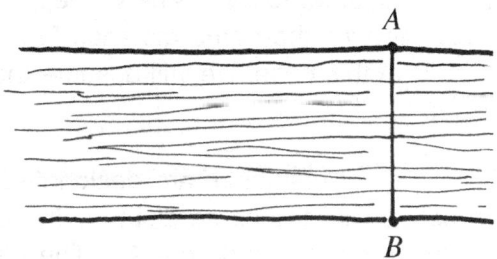

*Fig. 83: For Assignment 3*

4. With the help of a protractor and a linear measure, how can one determine the height of a tower, a house, or a tree from the length of the shadow cast by the sun?

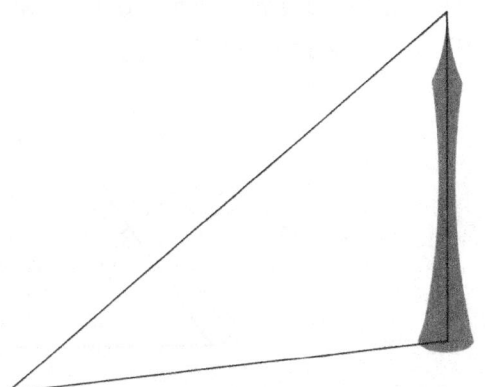

*Fig. 84: For Assignment 4*

# THE IMPORTANT LINES OF A TRIANGLE

## The Perpendicular Bisector of a Triangle

To introduce this topic, a story about a little situation can be told: Once there were three farmers and their families who lived a certain distance apart on a flat plain, such as on the Hungarian Pussta.[42] They decided to work together to build a new, deeper well. In order to be fair, they dug the well the same distance from each family. Where did they dig? (In this case, we will assume that the flow of water is not an issue.)

Solution: The places of residence are designated as points A, B, and C. All the points with the same distance from A and B are on the perpendicular bisector of side c = $\overline{AB}$. The well should be somewhere on this line. It should also be the same distance from A and C, that is, on the perpendicular bisector of side b = $\overline{AC}$, and on the perpendicular of a = $\overline{BC}$. We designate the perpendicular bisectors with $m_a$, $m_b$, and $m_c$, according to the side of the triangle to which they are perpendicular. A freehand sketch on the blackboard can make this clearer. Finally, the construction of the perpendicular bisectors of a triangle is drawn on the blackboard and then the children put it into their subject books.

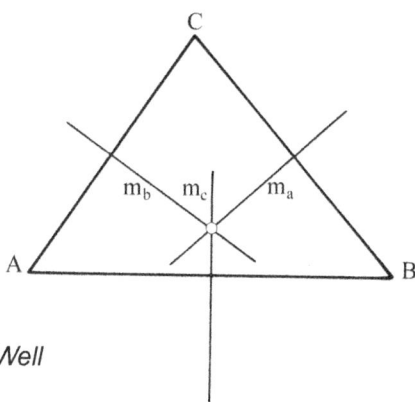

*Fig. 85: The Position of the Shared Well*

Since the children can, for the most part, freely choose the points A, B, and C (only they should not lie on a straight line), many varied situations come about. However, when the construction has been carefully done, all three perpendicular bisectors intersect at one point M. Is this necessarily the case? Yes, because all the points on $m_c$ are the same distance from A and B; all the points on $m_b$ the same distance from A and C; and all the points on $m_a$ the same distance from B and C. The point of intersection of $m_c$ and $m_b$ is equally distant from A, B, and C.

**The Circum-Circle**

Since M is the same distance from A, B, and C, then a circle around M that goes through A must also go through B and C. This circle surrounds the entire triangle so that the corners of the triangle are on the circle line. It is called the *circum-circle*.

To construct the middle point M, it is enough to construct *two* perpendicular bisectors and intersect them. The third can serve as a check for accuracy.

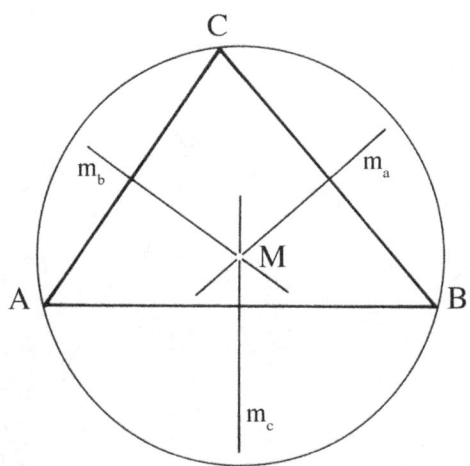

*Fig. 86: The Circum-Circle of a Triangle*

Assignment: Construct the intersection point M of the perpendicular bisectors for various isosceles triangles. Begin with a tall isosceles triangle, and let it change from acute-angled to equilateral and right-angled, and finally to obtuse-angled. How does the point M move?

We observe: If the apex falls to the base of the triangle, so also does the intersection point M of the perpendicular bisectors. With an isosceles right triangle, M lies right in the middle of the base. (Question: Does this also apply to right triangles that are not isosceles?) If the apex falls further, then M travels down, even outside of the triangle. In an obtuse-angled triangle the intersection point M of the perpendicular bisectors lies outside of the triangle surface.

This leads to a new and interesting question: If we look at Fig. 87 and revisit the well diggers, then we understand that M would be a *fair* solution for the placement of the well because it is equally distant from all three families. But, at the same time, this fairness is *impractical* because if M were closer to the triangle, the paths to the well would all be different, but *everyone* would have a shorter path. Those who continually harp on fairness must eventually accept that there may be enormous disadvantages. Sometimes, it is not the smartest position to hold.

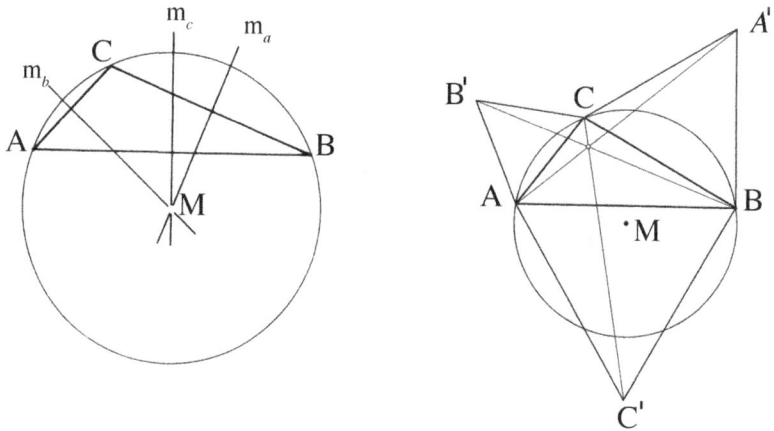

Figs. 87 and 88: Favorable and Unfavorable Placement for the Well

Question: Where must the well be dug so that the sum of *all of the paths* from the farms to the well will be the shortest making for the least amount of walking for the whole community? The solution is the so-called *Fermat* point F. To find the Fermat one constructs an equilateral triangle over each side of the original triangle—$\triangle ABC'$, $\triangle BCA'$, and $\triangle ACB'$. The point of intersection of $AA'$, $BB'$, and $CC'$ is the *Fermat* point F. [43]

*Assignments for the Perpendicular Bisector of a Triangle*

1. Begin with an equilateral triangle. Construct the perpendicular bisector and draw the circum-circle. Let the apex of the triangle move in such a way that the triangle always remains isosceles. How does the middle point of the circum-circle move? How does its radius change? Draw various examples. What happens when the apex comes very near to the base?

2. Begin the same as in assignment one, but this time shear the triangle sideways. Describe the changes of the circum-circle as in assignment one.

3. The circum-circle of a $\triangle ABC$ measures 5 cm. Construct $\triangle ABC$ when:

a) $a = 6.5$ cm       and       $b = 6.5$ cm

b) $a = 10$ cm        and       $\beta = 45°$

c) $a = 5$ cm         and       $\beta = 45°$

## Heights of a Triangle

When building scenery for a theater set, one must make sure that it can fit through the back door to the scenery storage area. If a triangle-shaped backdrop that is supposed to represent a cliff face, for instance, is brought from the stage to the backstage storage area, there are three sides upon which it can be moved. It has, in general, three different *heights*. In a triangle these heights are the lengths between an angle and the point where the perpendicular line has been dropped from this angle to its opposite side. As with the perpendicular bisector, one describes the heights according to the associated sides; that is, $h_a$ $h_b$ and $h_c$.

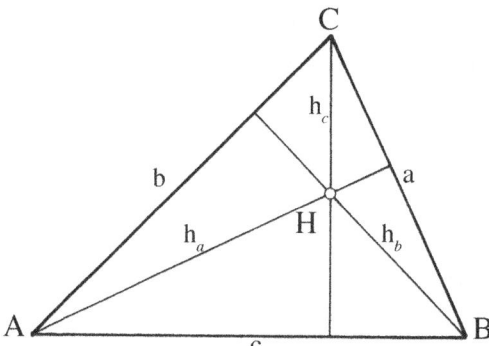

*Fig. 89: Heights of a Triangle*

In each case, the perpendicular bisector and height stands on a triangle side; $m_a$ and $h_a$, $m_b$ and $h_b$, and $m_c$ and $h_c$ are each *parallel*. When does a height coincide with the associated perpendicular bisector? When are all the pairs the same?

## Theorem of the Heights Intersection Point in a Triangle

The intersection of the three straight line segments determining the heights of a triangle is called the *height intersection point H*.

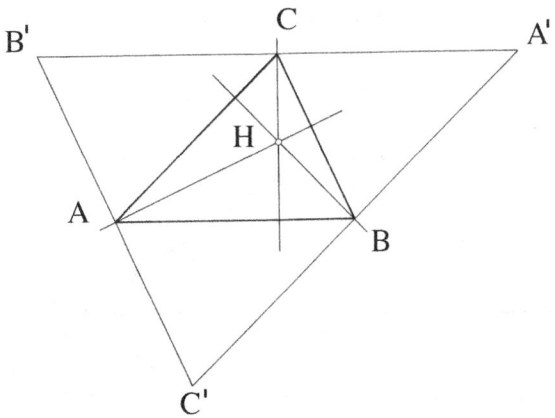

*Fig. 90: Proof that the Heights Intersect at One Point*

This can be understood if one supplements the triangle with other triangles of the same size and form, so that a new triangle △ABC is formed in which the height segments are the perpendicular bisectors. In the triangle △ABC, the perpendicular bisectors are $h_a$, $h_b$, and $h_c$. So, following the above principle, they intersect at one point.

Some of the more difficult but very satisfying triangle constructions are those in which the height as well as the sides and angle are given. However, one needs a minimum of three factors to construct a triangle. Assignments of this type require (and cultivate!) a systematic procedure as has been described since Greek antiquity as the Seven Steps. To explain the process here are three problems:

First Problem: This is known about △ABC: c = 8cm; a = 5cm; $h_c$ = 3cm. Construct the triangle!

Solution:
First step (*Protasis*): In general, this solution has to do with constructing a triangle for which one side, its associated height, and one other side are known. A freehand sketch should definitely be completed in order to get a clear idea of the geometric situation.

The given determined sizes do not have to be correctly drawn, merely estimated.

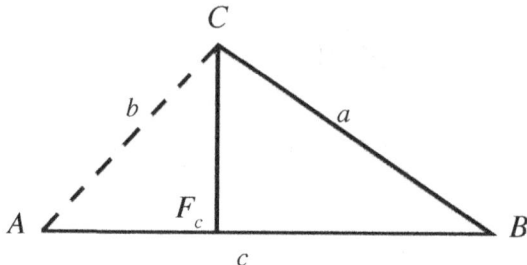

Fig. 91: Freehand Sketch

Using color is recommended; for example, known parts could be drawn in green and unknown parts in red. Here we have used solid lines for the knowns and dotted lines for the unknowns (Fig. 91).

Second step (*Ekthesis*): This problem stipulates certain sizes. In this way the general exercise has limitations. It is possible for such a problem to become unsolvable when, for example, in the construction of a triangle the sum of two sides is not always longer than the third side.

Third step (*Analysis*): Now we have grasped the situation clearly. We know all the necessary terms such as triangle, side, height, etc. We must now use our imagination, our inner, productive activity. The rules that guide this productive activity and protect us from fantastic combinations, must be firmly kept in mind. In light of this exercise, we proceed as follows: We can take side c as the starting point because the height h already stands in relationship to it, although h can be put many places on c, even onto the extension! But, h must always stand perpendicular to c.

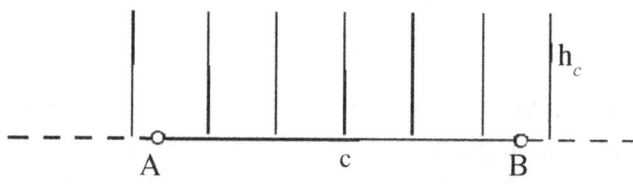

Fig. 92: Possible Positions of h

We know the length of a and a must join B and lead to the further endpoint C of $h_c$. The triangle consisting of a, $h_c$, and the segment of c between the height base points F and B, that is $\triangle F_c BC$, must always be a right angle at $F_c$. Such right angles, as we already know, always lie on a half circle, in this case above a. This brings us closer to the solution. We begin with a = 5 cm, draw a half circle over a and put an arc at c with the radius h that intersects the half circle at $F_c$. Line c must then go through B and $F_c$. With c = AB = 8 cm, we get A. With that we have determined all the angles of $\triangle ABC$.

Fourth step (*Apagoge*): $\triangle ABC$ actually does have the required characteristics. 5 cm is the length of a, c is 8 cm long, and $h_c$ is 3cm long. According to the third rule of congruence, the partial triangle $\triangle BCF_c$ is clearly determined by a, h, and the right angle. $\triangle ABC$, according to the second rule of congruence, namely by a, β, and c, is also clearly determined.

Fifth step (*Kataskeuae*): Now we can carry out the construction with full precision onto paper or the blackboard that which was previously only completed in our thoughts.

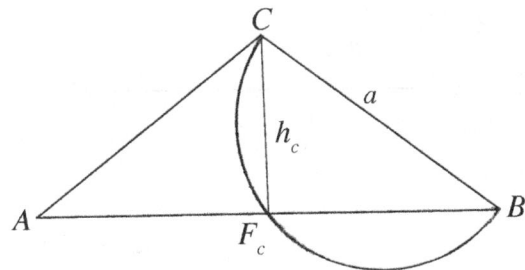

Fig. 93: *The Figure of the Construction*

Sixth step (*Apodeixis*): We make certain that the triangle has the required characteristics.

Seventh step (*Diorismos*): The problem is solvable only when a > $h_c$. The solution, except for mirroring, is clear.

Second Problem: It is known about △ABC that: b = 7 cm, $h_b$ = 4cm, α = 45°. Construct the triangle.

Solution:
Steps 1. and 2. One side, the associated height, and one adjacent angle are given.

Step 3. Line b can be drawn in any position with the end points A and C. α will be placed at one end (A). The second leg of α gives the direction of c. (The length of c is still unknown.) Line $h_b$ is perpendicular to b. (The exact position is unknown.) The second end point B for all possible positions of $h_b$ lies on a parallel to b at distance $h_b$ (on the side of b, where the free leg of a is). At the place where this parallel line segment intersects the free leg of α, is where B is. The triangle is now completely determined.

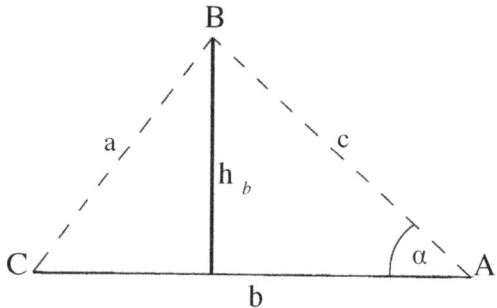

Fig. 94: Plane Figure

All the necessary steps are known: Making a line segment (b), joining an angle (α) at an endpoint of b, and constructing a parallel line segment to b at distance $h_b$. The actual construction can now take place.

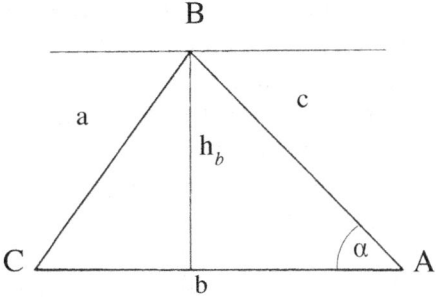

*Fig. 95: Construction of the Triangle*

The triangle has the required characteristics. There is only one solution because line c and the parallel line to b intersect at exactly one point.

Third Problem: This is known about △ABC: $h_b$, $h_c$ and $\alpha < 90°$. How can it be constructed?

Solution:
Steps one and two: This is a general problem since no fixed sizes for the three pieces are given. A plane figure will help clarify.

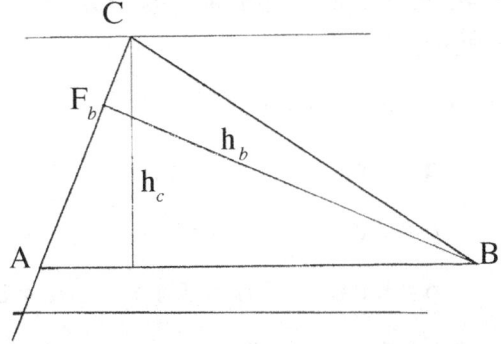

*Fig. 96: Plane Figure*

Third step: The triangle is divided into three triangular parts through the height. For example, we look at the triangle △$ABF_b$. One side ($h_b$) and two (therefore all three) angles are known. We can solve the problem in various ways. For example, we begin with $h_b$, join a right angle at end point $F_b$, maintaining the position of the line segment b, put $\alpha$ anywhere with b as a leg so that the line segment c is maintained and this, through the second end point of $h_b$ (B), positioned parallel. So, now we have $\triangle ABF_b$. C lies at distance $h_c$ parallel to AB and, naturally, on the line segment $AF_b = b$. With that $\triangle ABC$ is completely determined.

Fourth step: The triangle does possess the required characteristics.

Fifth step: An actual construction is impossible without given sizes.

Sixth step: ?

Seventh step: If $\alpha = 90°$, then $h_b$ and $h_c$ are legs of a right triangle.

*Exercises for Triangle Height*

1. Construct the height for variously defined triangles. Where is the intersection point for the heights in acute, obtuse, isosceles, and right triangles?

2. Construct triangle $\triangle ABC$:

| | | | |
|---|---|---|---|
| a) | a = 6 cm | c = 9 cm | $h_c$ = 4 cm |
| b) | b = 7 cm | a = 5 cm | $h_b$ = 4 cm |
| c) | b = 8 cm | $h_b$ = 5 cm | $\alpha = 60°$ |
| d) | a = 9 cm | $h_a$ = 5 cm | $\alpha = 45°$ |
| e) | $h_b$ = 7 cm | $h_c$ = 7 cm | $\alpha = 60°$ |
| f) | $h_a$ = 6 cm | $h_b$ = 8 cm | $\gamma = 60°$ |
| g) | $h_a$ = 7 cm | $h_b$ = 7 cm | $\gamma = 90°$ |

## The Angle Bisectors in a Triangle

When we look out, a world of color appears before us. One speaks of a person's *field of vision*. Different people enjoy different widths of this field of vision and, in general, it becomes narrower with age or the wearing of glasses. If we want to look at something in very exact detail, then we turn our attention and *vision* to it. This ray of attention lies just about in the middle of our spatial angle of vision. It halves all the angles between the outer limits of vision.

To demonstrate, one can ask three children to stand in a triangle. Each child should look in the middle between both lines of sight to the other.

In the middle, the teacher holds a vertical copper rod and lets the three children each direct her in such a way that the rod can be seen in the middle direction between the other two children. One eye should be closed. The children's vision goes in the direction of the bisecting line of an angle of the triangle. The situation can be illustrated with a freehand figure on the blackboard.

Now, the children can construct bisecting lines of angles for different triangle forms, which are designated with $w_\alpha$, $w_\beta$ and $w_\gamma$. The three bisecting lines of the angles in a triangle also intersect at one point, W.

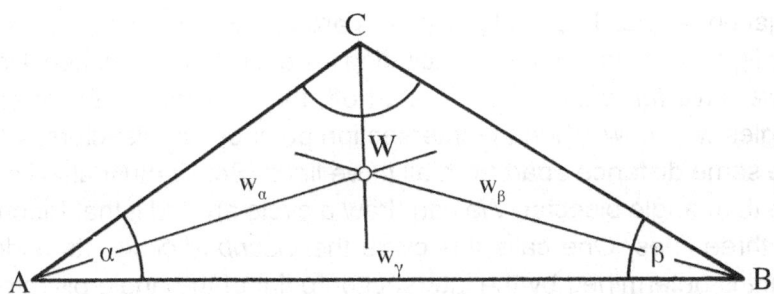

*Fig. 97: The Intersection Point of the Bisecting Lines of the Angles*

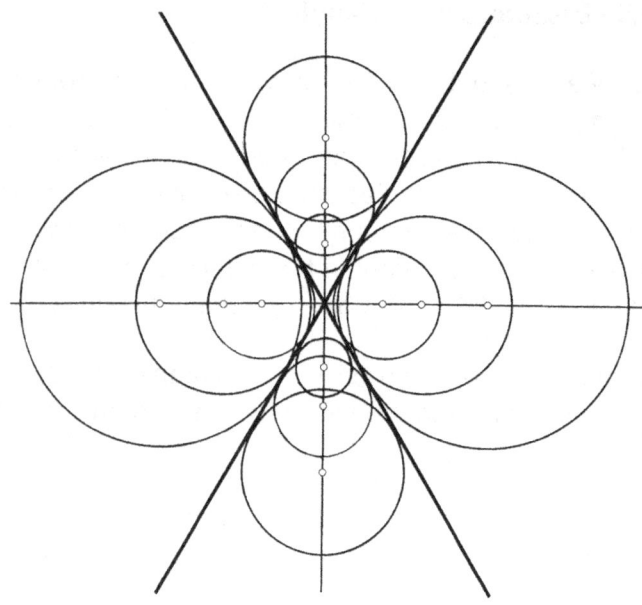

*Fig. 98: Tangent Circles of Two Straight Lines*

**The Inscribed-Circle**

In order to understand the special characteristics of the intersection point of the angle bisectors, we look first at only *two* lines.[44] The middle points of circles which *touch* the two lines are lying upon the angle bisectors. Other circles, which also touch the two lines, have their middle points on the angle bisector of the adjacent angle. The angle bisectors are the two symmetry axes for the figure from the two lines, including the circle that is touched. Now, if we have *three* lines and, to start off, the bisectors of the interior angles $w_\alpha$, $w_\beta$, $w_\gamma$, then the intersection point of two bisectors, W, is the same distance apart from all three lines. So, W must also lie on the third angle bisector. We can draw a circle around it that touches all three lines. One calls this circle the *inscribed-circle*. Its middle point is determined by the intersection point of the angle bisectors. The exact radius can be found by dropping a perpendicular from W onto one of the sides. The length of the perpendicular is the radius of the inscribed-circle.

## The Principle of the Intersection Point of the Bisecting Lines of Angles in a Triangle

The three bisecting lines of the angles in a triangle intersect at one point, W. It is the middle point of the inscribed-circle.

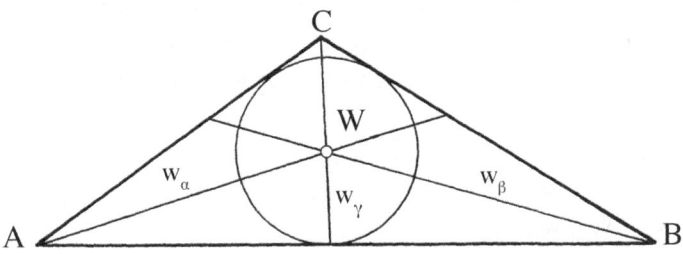

Fig. 99: The Inscribed-Circle

This rule does not generally apply to quadrangles, as every rectangle shows, that is not at the same time a square. However, in the next volume of this series we will study such special quadrangles that have an inscribed-circle (tangent squares).

*Exercises for the Bisecting Lines of the Angles in a Triangle*

1. Construct the inscribed-circle for different triangles.

2. Draw a horizontal line c. Choose two points A and B that are 8cm apart on the line. Construct an isosceles, right-angled triangle over $\overline{AB}$. Construct the inscribed-circle. Now rotate the lines $\overline{AC} = b$ and $\overline{BC} = a$, in 15° increments counterclockwise so that C travels toward the top. Construct the inscribed-circle for every increment. Rotate b and c a total of 90° (180°? 360°?).[45]

3. Construct a regular, six-pointed star in a circle and the inscribed-circle within it.

4. Construct △ ABC from:

| | | |
|---|---|---|
| c = 7 cm | α = 60° | $w_\alpha$ = 5 cm |
| c = 6 cm | β = 50° | $w_\beta$ = 6 cm |
| a = 8 cm | γ = 90° | $w_\gamma$ = 5.6 cm |

## The Adjacent Circles of a Triangle

As we remember, two lines have two collective angle bisectors. Let us look at the full line (*three-sided figure*) of a triangle. There are actually *six* angle bisectors; that is, three for the interior angles, and three for the exterior angles. We designate them as: $w_\alpha$, $w_{\alpha'}$, $w_\beta$, $w_{\beta'}$, $w_\gamma$, $w_{\gamma'}$. In each case, the following intersect in one point:

$w_\alpha$, $w_\beta$, $w_\gamma$
$w_{\alpha'}$, $w_{\gamma'}$, $w_\beta$
$w_{\beta'}$, $w_\alpha$, $w_{\gamma'}$
$w_{\gamma'}$, $w_\alpha$, $w_{\beta'}$

These points are all the middle points of circles that are touched by all three lines. If one is using the angle bisector of an exterior angle, one speaks of an *adjacent circle*. A three-sided figure has one inscribed-circle and three adjacent circles.

The angle bisectors themselves of the exterior angles form another triangle in which the angle bisectors of the interior angles are the height lines.

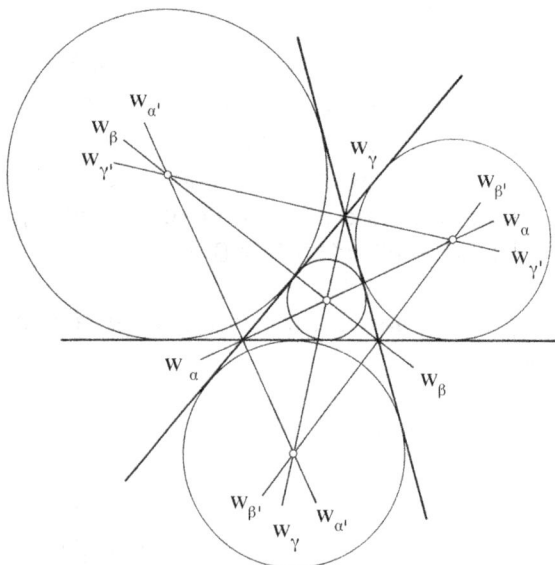

*Fig. 100: The Angle Bisectors of the Interior Angles, Exterior Angles, and the Adjacent Circles*

# The Platonic Solids

One of the most beautiful subjects for elementary, but also higher, geometry is the Platonic Solids. Named after Plato (427–347 BC) because he first presented them in his *Timaeus*, they may have been known in earlier times.[46] Euclid, who was responsible for the first great work on geometry around 300 BC and which has been in use for over two millennia, crowned his work with the Platonic Solids. Johannes Kepler, the great astronomer (1571–1630), became famous for his written work *Mysterium Cosmographicum*. In that book he looked at the Platonic Solids in relationship to our solar system, though in his earlier work still only in respect to orbits. There is much more to be learned from these solids.

*Tetrahedron*   *Cube*   *Octahedron*

*Icosahedron*   *Dodecahedron*

Figs. 101–105: The Platonic Solids

To my knowledge, Rudolf Steiner did not give any specific advice about teaching the Platonic Solids. But he did point out a fundamental aspect of geometry: Feelings and thoughts are swimming in human speech. "And the human being would be thrown into a certain chaos of life already through speech, and other things as well, if he did not get the stability that comes from mathematics. Those who look deeper into life know how many people are protected from neurasthenia, hysteria, or worse, only by having learned to look at triangles, quadrangles, tetrahedrons, etc. in the right way."[47] What many people experience as the beauty of these solids may have something to do with this. Ultimately, because of their beauty and wonderful adherence to geometric laws, this topic is worth discussing thoroughly.

One place this study can easily be incorporated is in the study of regular polygons. We have already mentioned the tetrahedron (four-sided). The other Platonic Solids can be easily included. If there is time and opportunity, the best way is to model them from a ball of clay. Depending on which direction the compressing and stretching energies act upon the clay, the various forms are produced.

*The cube or hexahedron* (hexahedron = six-sided) comes about when three spatial directions standing perpendicular to one another each presses a pair of equal counter forces to the middle point.

If the forces act in the direction of the four spatial diagonals of the cube, that is, from the eight corners, then an octahedron (eight-sided) is formed with its eight triangular surfaces. Conversely, one gets the cube when one presses the octahedron from the opposing corners. This illustrates the close relationship between these two solids, as is also found in the crystal forms of pyrite, for example. This activity can be used as a bridge to the study of crystal forms from geometry as suggested by Steiner for the sixth grade.

A similar sibling-like relationship to the cube and the octahedron exists with two higher solids: the *dodecahedron* (twelve-sided) and the *icosahedron* (twenty-sided). However, to be able to freehand sculpt from a ball of clay requires much skill.

*Fig.106: Pyrite Crystals*

## The Language of Geometric Forms

For a more intellectual discussion, one can talk with the students about the sentient impressions made on us by the wonderful harmony of these solids. In his work *Timaeus*, Plato connected these five solids with the elements: the tetrahedron with fire, the octahedron with air, the icosahedron with water, and the cube with earth. The dodecahedron was labeled the quintessential form in the world of etheric formative forces.[48]

Just as the individual solids leave their characteristic impressions, they can also have very different effects in different positions. The cube especially, in the successive positions of standing flat, on the edge, and on the corner, shows a certain loosening from gravity as would be comparable to experiencing solid, liquid, and airy elements. Paul Schatz referenced this vividly.[49]

*Fig. 107: Cube in Three Positions*

## A Beginning of Conceptual Understanding

Parallel with constructing the solids, the students should be led through thoughtful considerations at the beginning of each lesson in which the laws and principles are presented. One can start from the following questions: What surface forms contained within each individual solid? How many surfaces, edges, and corners do they have? How many surfaces and edges meet in one corner? How many edges and corners surround one surface? What symmetries do these solids have? And many other questions.

From the results of the discussion we can summarize the following: Three solids are formed by triangles (tetrahedron, octahedron, icosahedron), and one each from quadrangles (cube), and pentagons (dodecahedron).

The following table lists the number of corners, edges, and surfaces of each solid. One can recognize the sibling-like pairing of the cube and the octahedron as well as the dodecahedron and the icosahedron: The one solid has as many corners as the other has surfaces while the number of edges are the same. With one, a surface is surrounded by as many edges as go through one corner with the other. The tetrahedron is, so to say, androgynous: It has as many surfaces as corners.

|              | Corners | Edges | Surfaces |
|--------------|---------|-------|----------|
| Tetrahedron  | 4       | 6     | 4        |
| Cube         | 8       | 12    | 6        |
| Octahedron   | 6       | 12    | 8        |
| Icosahedron  | 12      | 30    | 20       |
| Dodecahedron | 20      | 30    | 12       |

In order to form one corner, at least three surfaces must be pushed together, and between them, at least three edges must meet in the corner. One surface must have at least three corners (points), and between them three edges; that is, a triangle. Here, the principle of polarity in projective geometry is illuminated, as Steiner suggested for further study in the tenth grade.[50]

**There Are Only Five Regular Solids**

The complete acceptance that there are only five regular solids is not easily achieved at this level.[51] However, we can take a few first steps in that direction.

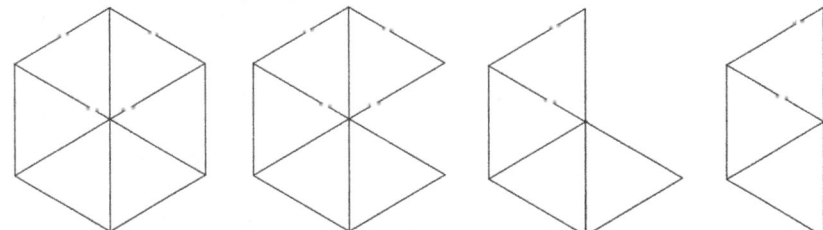

*Fig.108: Forming Corners out of Triangles*

The solids that are formed out of triangles (icosahedron, octahedron, tetrahedron): The simplest way to geometrically form the Platonic Solids is to start with a regular hexagon which has been formed in the familiar way out of a circle by drawing six chords from the radius. If we connect the opposite points, then we get six equilateral triangles. It is not possible to form a three-dimensional angle or corner from this because it is a plane, or two-dimensional, surface.[52] In order to form a three-dimensional form, we must remove at least one triangle. If we connect both of the freed edges we can make a roof out of five triangles. This is part of a icosahedron. Naturally, in order to form a complete solid, more triangles must be attached. We will discuss that later.

If we take two triangles out of the hexagon, then we can produce the empty corner of an octahedron out of the remaining triangles. Two three-dimensional forms of this type make a complete octahedron.

If we remove three triangles from the hexagon, then only three remain which is the lowest number of triangles possible in order to get a three-dimensional form. If we think of them as being put together, then one triangular surface remains open. For this reason the tetrahedron is closed by four equal triangles.

The cube: Assemble four equal squares around a point. With their four interior angles of 90° each, they make a full angle, 4 • 90° = 360°. In order to create a three-dimensional form, one square must be removed. The three remaining squares form one cube angle. We need two such angles to get the complete cube. No three-dimensional form can be made with less than three squares.

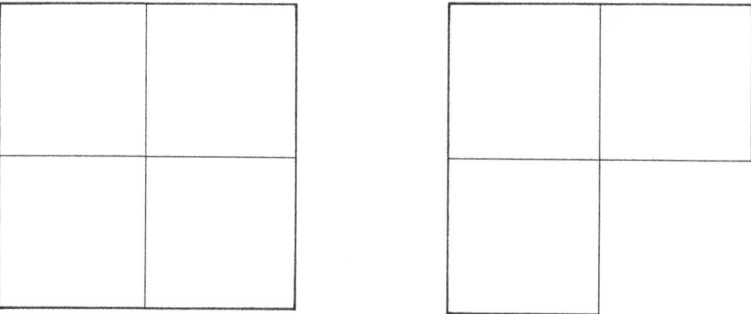

*Figs. 109 and 110: Constructing a Cube from Squares*

The dodecahedron: The regular pentagon has interior angles of 108°. Therefore, three pentagons pushed together on one point, with their interior angles of 108° = 324°. There remains a space of 36°. For this reason an empty corner can be formed: The angle of a dodecahedron (see Fig. 116).

No empty corners can be formed from hexagons or other polygons because three hexagons with their interior angles make 360°, that is, a plane figure. This is the simplest understanding that there can be only five forms as previously discussed.

For a strong proof of the existence of exactly five solids it is necessary to consider a few more things that space in this volume does not allow. It is especially important to show that attaching further equal surfaces onto the open edges will again form empty corners and these, finally, will form enclosed solids.

## Constructing the Platonic Solids

In order to produce the Platonic Solids we must first construct grids. This is like unwrapping a solid onto the surface from which the solid can be reconstructed using folds and glue. For the dodecahedron and icosahedron, it is recommended that one make two separate grids. In order to develop the necessary skill we begin with the tetrahedron and follow with the cube and the octahedron. The tabs to be glued should be drawn before cutting is begun. They should be large enough and slanted so that they do not interfere with other side surfaces once they are bent. Before folding the edges should be creased toward the inside.

• The Tetrahedron: We begin with its construction as has already been shown in Fig. 17. Naturally, one must first decide where the glue tabs should be before cutting out and which edges will be glued together. For every edge of the solid there should be one glue tab.

• The Cube: With the cube one can ask an interesting question that should stimulate the imagination of the students: The six squares of the cube can be drawn in a connecting grid. How many variations are possible?

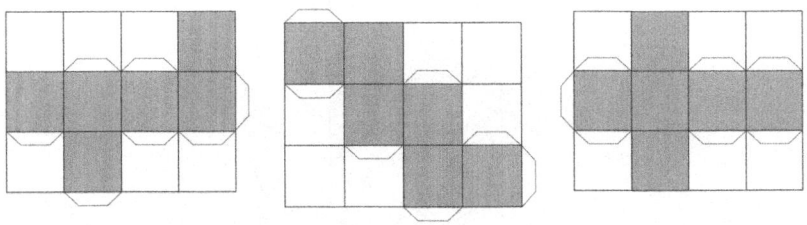

*Figs. 111–113: Three Variations for the Cube Grid*

For assistance one can copy a larger rectangle that is divided into 3 x 4 = 12 equal squares (Fig. 111). It takes six of them to form a cube. There may not be more than four in a row (thus the limitation of a 3 x 4 square), and no more than three can come together in one corner. However, when they are glued together three squares must come together at every corner. That is why one must consider well which edges meet when the grid is folded. It is a help to designate the edges that go together by marking them with corresponding numerals. Figs. 111 through 113 are a few examples. The activity of finding new grids and putting the right edges together trains active spatial thinking and is of great benefit to the students at this age.

- The Icosahedron: We begin with two large, equal circles, as large as the compass and the cardboard or paper will allow. (The compass may need to be adjusted to a larger span, or one that is already long enough can be passed around among the students.) The six-times repeated radius gives the fundamental structure. The diagonals form divisions as can be seen in Fig. 114. If two adjacent triangles are removed, ten triangles are left which fold to make half an icosahedron. Both halves glued together make the complete solid. Do not forget the glue tabs!

Fig.115 shows a grid for assembling the complete icosahedron from one piece. This grid is easy to construct using the circle field or rosettes.

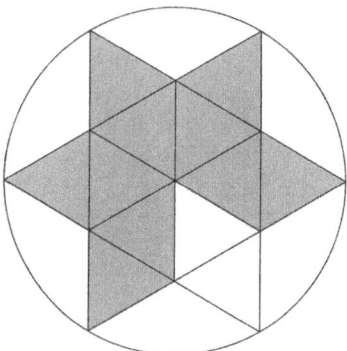

*Fig. 114: From Hexagon to Icosahedron*

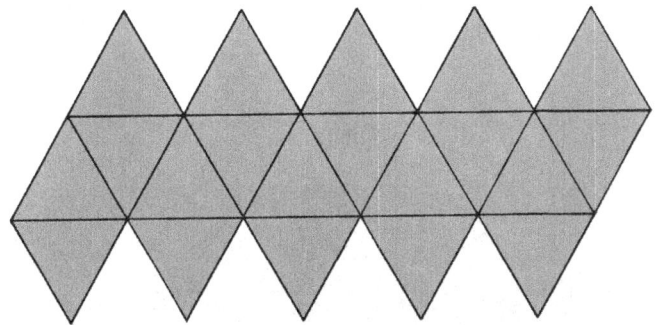

*Fig. 115: An Icosahedron*

- The Pentagonal Dodecahedron: In this case we also begin with as large a circle as possible that we also draw twice. In one of them we construct the pentagon as described in Volume 2.[53] One must be careful that the five sides of the pentagon lead back to the starting point. Even small discrepancies must be corrected by slightly repositioning the compass, or the pieces will not come together properly. If the first pentagon is drawn onto the circle with enough exactitude, then it will also easily copy onto the second circle. Now the pentagon and the pentagram will be drawn. In the center another pentagon will be formed. We also put in the pentagram in it, thereby getting a grid for half a pentagonal dodecahedron. When both parts are glued together using tabs we have the complete solid.

If the teacher and the class are especially interested in these wonderful solids, they can also try to make outline models. If one uses a hard, quick-drying model glue, one can also use straw, thin dowels or the stalks of certain dried grasses to model the Platonic Solids.

Of course, all of these various projects are only the very first steps on the subject of the Platonic Solids. They are significant in many areas of life. There are always new and amazing connections to be found when one delves deeper into the subject. The students should take with them the impression they have only touched on the very beginning of the laws and principles governing these solids, and that their significance goes much further than the students are able to realize at this stage.[54]

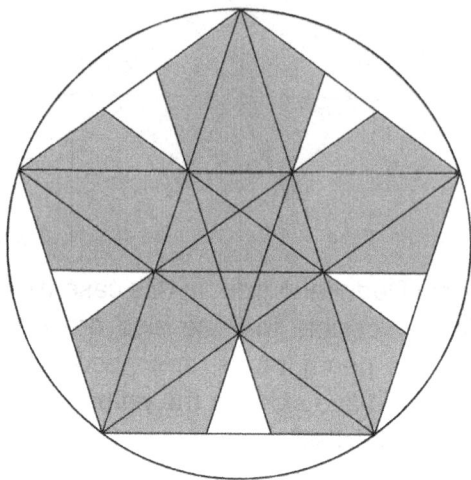

Fig. 116: The Pentagonal Dodecahedron

## Simple Projections

Steiner repeatedly and emphatically pointed out the significance of projective geometry in education, even for the earlier grade levels which have traditionally not dealt with such subjects. For this reason I begin with some of the things he had to say.

In the second lecture Steiner gave to the teachers at the founding of the first Waldorf School in 1919, he spoke, among other things, about projection and shadow and how this subject should be taught in the sixth through eighth grades: "Then in the sixth class introduce simple lessons in projection and shadow that are done using freehand drawing as well as with a ruler and compass, and so on. See to it that the child gets a good understanding and can trace and copy well, such as when here is a cylinder, here a sphere, and the sphere is illuminated by the light, what does the shadow from the sphere onto the cylinder look like. How shadows are cast! So, simple projection and shadow lessons should appear in the sixth school year.

The child must get an idea of, and must be able to imitate, how shadows are cast onto flat areas, bent areas, other areas more or less flat, and physical objects. In this sixth school year the child must get an understanding of how the technical connects to the beautiful, how a chair can be technically suited for a purpose and at the same time, incidentally, have a beautiful form. And the child should go into this understanding of the connection between the technical and the beautiful.

Then in the seventh school year everything should be nurtured that has to do with penetration. As a simple example, the teacher says: "Here we have a cylinder that is penetrated by a post. The post must pierce through the cylinder." The teacher must show what kind of intersection there is in the cylinder when it is pierced and when it comes out the other side. It must be learned with the child, something like this: What happens when bodies or areas penetrate each other? What is the difference if a stove pipe above goes perpendicular through the ceiling, whereby the penetration is circular in form, or slanted, whereby the penetration is elliptical in form? Then, in this year, the child must be furnished with a good understanding of perspective. Simple perspective drawing, shortening in the distance, lengthening up close, overlapping, etc. And then again the connection of the technical with the beautiful, so that one calls forth from the child a notion of whether it is beautiful or not. If ever, let us say, the partial covering of a wall of a house was effected by a ledge, such a ledge can look nice or not so nice as a covering for the wall. Such things have an immense effect when they are taught to the child in the seventh school year, that is, when he/she is thirteen or fourteen years old. This is all raised into the artistic as one goes into the eighth school year.[55]

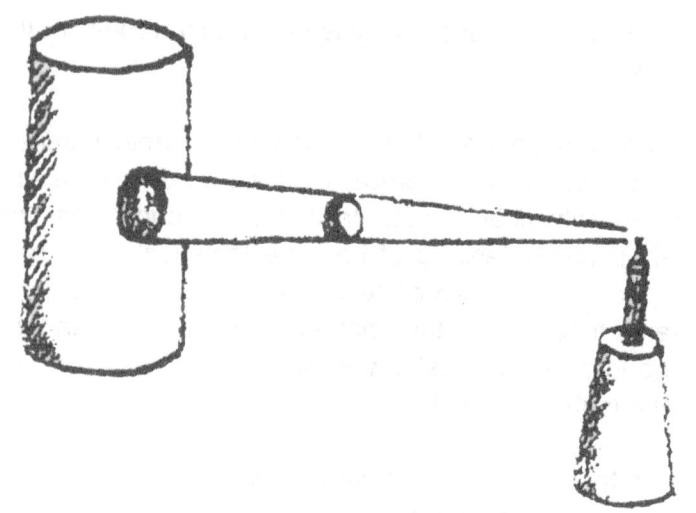

*Fig. 117: Sketch by Rudolf Steiner about Projection and Shadow*

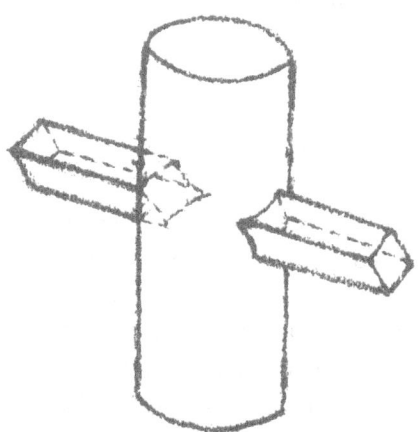

*Fig. 118: Sketch by Rudolf Steiner involving a solid penetrating another solid*

In a 1920 lecture for teachers in Basel, Switzerland, Steiner spoke in general terms about the educational significance of a flexible perception of space, about the importance of also being responsive to the inner being during a spatial process, and about the possibilities there are in active games. He went on to say:

"Then, however, it is of very special importance now, really out of that, what was observed in this way, to transition into remembering what was observed. Namely, it is for the development of spatial perception ... of great importance, when I let shadows be cast from different curved areas by bodies with different curvatures, and now try to call forth an understanding for the special configuration of the shadows. One can virtually claim that: When a child is capable of understanding why, under certain conditions, a sphere casts an elliptical shadow—that is something which can be grasped by a child already from age nine—then this ability to put one's self into the formation of areas in space has an enormous effect on the child's entire inner flexibility of perception and imagination. For this reason, one should see the development of the spatial sense as something necessary in school."[56]

At a conference for the teachers of the Waldorf School in Stuttgart, Germany, on January 16, 1921, a colleague asked Steiner about the more concrete presentation of lesson plan suggestions given in 1919: In teaching projections and shadows in class six, is it better to start from the point of view of art or geometry?

Steiner replied: The best thing, circumstances permitting, is something that forms a bridge between a mere geometry lesson and one that will lead to art. I don't think you can deal with it artistically. What is meant is the conic section. I would think the children really ought to know what the shadow of a cone is like in a given plane, so that they visualize it.

To a further question: Ought we to use expressions like "light rays" and "shadow rays"?, Steiner replied: That is a more general question. It isn't a good thing in projective geometry to use things that don't exist. Light rays don't exist and shadow rays exist even less. It isn't necessary to work with these concepts when teaching projections. You ought to work with spatial concepts. There are no such things as light rays and shadow rays, but there are cylinders and cones. And there is a shadow body which arises if I have a cone that is oblique and that is illuminated from a point that casts a shadow on an inclined plane. Then I have a shadow body that is really there. The child certainly should understand this shadow body as such, the boundary of the curves of the shadow body, just as later on in projective geometry he has to understand how one cylinder bisects another that has a smaller diameter. It is tremendously useful to teach the children this. It does not lead away from the artistic sphere but keeps them in it. It makes their thinking flexible. You can think flexibly if you know from the outset what kind of intersecting curve will arise when cylinders bisect. It is very important to give them things like this and not abstractions.[57]

Caroline von Heydebrand wrote a very concise formulation for teaching projection and shadow in the sixth and seventh grades in her work about the Waldorf curriculum. Sixth grade: "Simple projections and shadow drawing should now be introduced. The form and figure which lie in the shadow should be studied without paying attention to the construction, and the children should draw freehand as well as with ruler and compass. The teacher should seek to awaken in the child an understanding for the connection between the technical and the beautiful in the making of common objects."

Seventh grade: "Perspective is practiced including interpenetrating and overlapping of objects as well as foreshortening. An aesthetic feeling for functionally designed industrial objects should be awakened."[58]

That is about as far as it goes on specific information relating to teaching projection and shadow in the middle school. Converting this information into a lesson plan naturally requires a design that moves in the direction of the examples presented, such as the shadow of a sphere on a cylinder. Obviously, this does not have to do with a lesson in pure drawing where one attempts to recreate what is seen, but rather on the observance of spatial relationships that should be made transparent to the students to a certain degree without taking anything away from the construction work that takes place in the upper grades.

Creating shadow in the sixth grade can be seen as a preparation for drawing penetrated geometric forms in the seventh grade because the visible shadow, as has already been mentioned,[59] is to be understood as a cut surface, a cross-section intersecting plane of a shadow space with a surface area on which the shadow appears.

Today the question arises as to the significance of education in the area of spatial-related intelligence in light of software available for architects, engineers, and others with careers for which knowledge of geometry is essential in their work. Certainly not to replace computer-aided design systems or something similar, but rather in order to be able to interact with them in reality on a creative level! Similarly, having an automobile does not mean that motor development in children is unnecessary. Just the opposite, it means that motor development becomes much more important. In the same way, the use of calculators actually reinforces the ability to calculate and evaluate results. Thus, a not-merely mechanical application of software must presuppose the creative ability to interact with spatial forms. That is what makes the notes on transitioning from the geometrical to the artistic so very applicable in modern times—because often the machine can perform purely technical tasks with more accuracy than the human being, but the human being must understand what is behind the function for it to have relevance and meaning in the world.

# Teaching Projection and Shadow Drawing
## in the Upper Elementary Grades

The introduction of this part of the curriculum, which is best included with the light and shadow portion of physics in the sixth grade, should be connected with observations made in the fourth or fifth grade, as suggested in Volume 2 of this series. However, the students can now grasp spatial relationships with more willful forces. If these exercises have not been done earlier, then one can begin them immediately.

A nice beginning would be using a styrofoam or clay ball with a thread on it attached with a pin, for example, so it can be held in the sunshine without one's hand getting mixed in with the cast shadow.[60]

Neither the sunlight-flooded light space nor the forming shadow space behind the ball can be perceived. They exist, so to say, only as potential to create brightness and darkness on a material surface. What can be observed is the bright side of the ball with the sun shining on it and its darker side which holds the object's shadow.

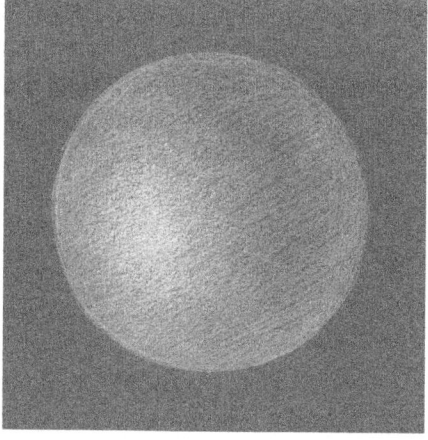

*Fig. 119: Lightness and Darkness on a Ball (E. Wannert)*

That we may speak of a shadow space or shadow body shows that if we put a flat, white surface (paper), that we will call the image area, into this shadow space, we can observe a differentiation, a sun-brightened light surface and a dark shadow, the cast shadow of the ball. If we move the image area parallel, then we can observe that the possibility (potential) of shadow formation is spatially stretched and geometrically structured. Behind the ball there is a somewhat cylindrically formed shadow space or shadow body surrounded by light space (Fig. 120).

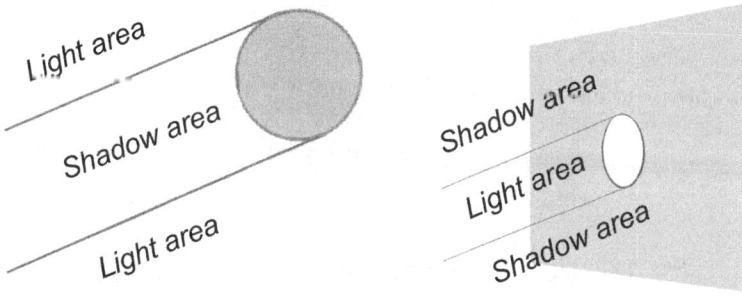

*Figs. 120 and 121: Light and Shadow Spaces behind a Ball and a Screen*

If we turn everything around by cutting a round hole in a larger piece of cardboard, then, when the paper is held perpendicular to the sun, we get a circular, cylindrical light space that is surrounded and bordered by a shadow space (Figs. 121 and 122). This entire formation of light and shadow spaces is especially significant because it calls upon imaginative thought in order to understand a spatial whole.

The following steps outline using the shadow of a sphere, but they could also apply to the circular hole in the screen.

*Fig. 122: Light and Shadow Spaces behind a Screen*

First, we call attention to how the shadow cast behind the ball is circular when we hold the image area, which makes the shadow appear perpendicular to the light. If we now turn this image area so that the cylindrical shadow space is cut at an angle, then elliptical shadow forms are created. Already in the fourth or fifth grade we have tried to make this understandable by remembering that a biased-cut cylinder (a salami for example) results in elliptical slices.

It is recommended that elliptical forms be practiced by doing a variety of freehand drawings because, all too often, the ends drawn are pointed instead of round. One can also call attention to where we see elliptical forms—namely, everywhere we see a circular form at an angle: the rim of a cup, a round lamp shade, or the rim of a plate.

*Fig. 123: The Elliptical Shadow of a Sphere (E. Wannert)*

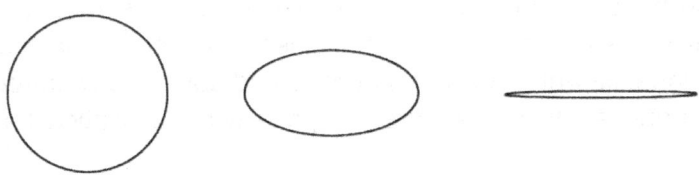

*Fig. 124: The Circle Seen at an Angle Becomes an Ellipse*

*Fig. 125: The Cast Shadow of a Sphere (E. Wannert)*

If there is a large ring available (something like a gymnastics ring), one can hold it upright in front of the class and let them draw the circle in the air with their hand. If one then tilts the ring little by little the children must draw ever more elliptical shapes until the shape finally becomes a line (Fig. 124). For perspective drawing it pays to thoroughly work out these elliptical forms.[61]

If we set the ball on the image area, then careful observance will reveal amazingly rich differentiations: The image area shines back in the shadow body and brightens it. Through this it can again shine into the cast shadow, that, in any case, does not appear to us to be evenly dark. Usually it appears to be lighter in the middle than on the edges. After these experiments and their careful observation and description, what has been seen can be drawn on paper using drawing charcoal or some other medium. The students will need plenty of time to complete each drawing. The next day the pictures can be hung up, compared, and especially successful efforts duly noted: Did someone observe the sheen of the sun on the ball? Were the different levels of brightness and darkness captured in all their nuances? Where are there especially good elliptical forms?

Again and again one should point out the effects of shadow formation in practical life. For example, it plays a big role in architecture. If a tall house is to be built next to a kindergarten, then the architects must check to see where the shadows appear at various times of the day and during all seasons in order to see how many hours of sunshine the kindergarten building will get. What plants can grow in the garden or not depends upon the shadows cast by the trees and the plants themselves! Interestingly, the variety of plant species to be found on the edge of a wooded area, with its changing amount of light and shade, is significantly larger than in the inner part of the woods where there is an equal amount of shade or in a sunny meadow where there is an equal amount of sunlight. It is the changing play of light and shadow that stimulates the variety of life.

And now to get to other geometric forms, thereby making possible a larger variety of phenomena as well as more individual work for the students. Every student will get a clay ball about ten centimeters in diameter that has been prepared the day before and kept moist with a wet towel. About one-third of the ball will be for forming as perfect a sphere as possible, another one-third for a cylinder, and the rest for a cone. When these geometric forms have been finished to complete satisfaction, every student will look for an appropriate light source (depending on the situation this could be the sun, a table lamp, a candle, or even a prepared flashlight bulb) and draws his or her clay forms with their corresponding shadows. At first, this can be done by just eyeing it, but in many cases this will not lead to very satisfactory results. There can be problems with the sculptural representation of the form as well as the exact shape and position of the shadow. When all are finished the pictures are hung up and preliminarily viewed. Various questions are given to the students to think about for the next day concerning specific conditions for the drawings, such as: How are the shadows connected to the forms? What form do they have with the various lights? How does one draw the forms so that they appear three-dimensional?

The next day one can discuss the questions given the day before. For a good three-dimensional representation of the forms we point out the differentiated divisions of light and dark, as the case may be. When we think of a ball with the sun shining on it, we can see that not all of the lighted surfaces have the same brightness. It is brightest where the surface is perpendicular to the light. If the surface on the ball is moved away from the light then it becomes darker. This can be seen on many of the drawings reproduced here. Geometric relationships also play a role here. However, that is not the only thing. The darkening up to the shadow on the ball (or other form) very much depends on the surface structure. The luster of a metal ball mirrors the surrounding space, often in sharp, bright lights. A clay ball has a mild, more equal glow, etc. One can actually make visible something of the characteristics of the material just by observing the light-dark relationships. In order to help the students understand the form of the shadow, we must bring in geometry.

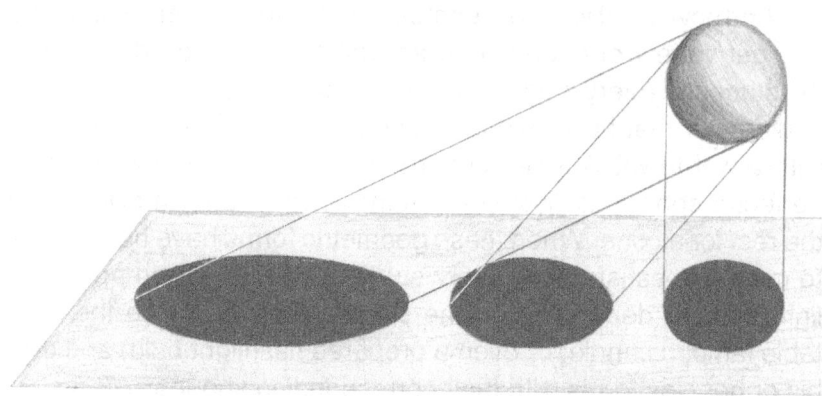

*Fig. 126: Elliptical Shadows of a Sphere with Different Lighting Directions*

*Fig. 127: Lighting from Behind and Above*

We already know that the shadow of a sphere on a horizontal plane with angled light is elliptical. With a geometric sketch we can make clear the shadow length according to the position of the image area to the light. If the light shines perpendicularly onto the image area, then the shadow is circular. The more angled the light is, the more the shadow is pulled into an elliptical shape (Fig. 126).

If the light comes from behind (Fig. 127) instead of angled from the side, then the shadow in the drawing is shortened accordingly. This construction, in all its details, is, without question, manageable for a sixth grade class. Naturally, the ellipse, as seen from the side, becomes smaller in a perspective representation.

If the light source is nearby instead of the far-away sun, then the shadow becomes larger accordingly. One difficulty for the students is that the real shadow relationships as they are observed in the experiments are, thinking correctly, actually corrected by us. When we draw, the distortion that occurs because of the position of the observer must be taken into consideration.

For all cast shadows of a geometric form, one can speak of an effective cross section that is defining for the shadow. It is a surface that would effect the same shadow as the form itself. With a ball this would be a circular surface perpendicular to the light. In an angled view the circular surface appears as an ellipse, its width also the width of the shadow.

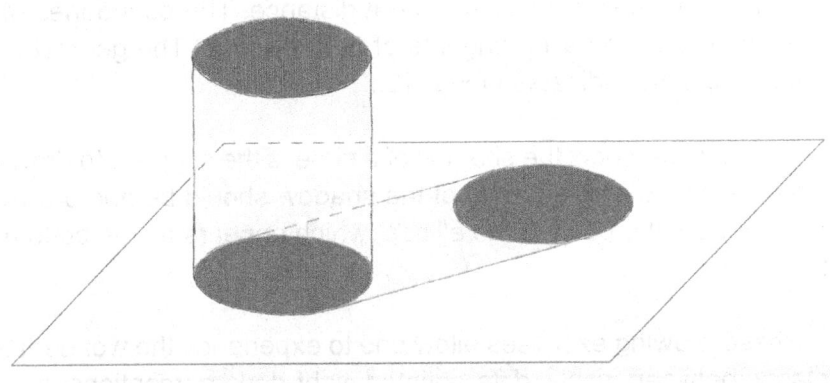

*Fig. 128: Construction of a Cylinder Shadow with Parallel Projection*

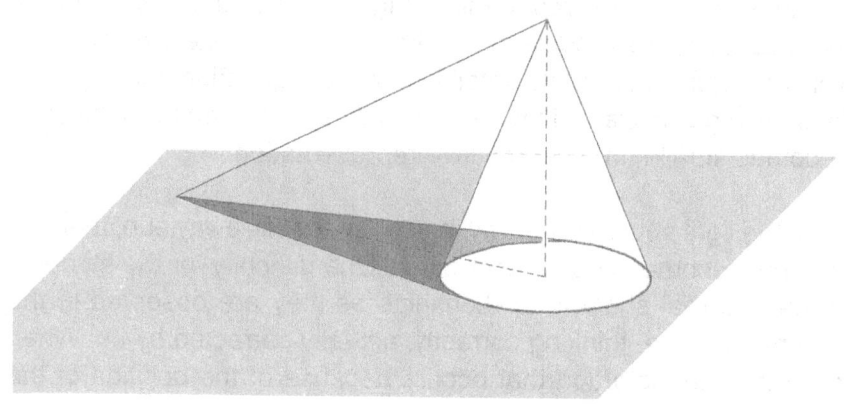

*Fig. 129: Construction of a Conic Shadow*

Generally, the circular bottom and circular top of a cylinder will also appear as ellipses. If the cylinder is standing upright, then, in pure geometric terms, the bottom, on itself, and the top display equal ellipses in a parallel position if we have parallel projection in which the light is shining from a great distance. The boundaries of the shadows are parallel tangents of both ellipses. The geometric relationships are indicated in Fig. 128.

In certain respects the shadow of a cone is the simplest to draw: One determines where the tip of the shadow should be and draws tangents from the point to the ellipse, which appears as the bottom of the cone.

These drawing exercises allow one to experience the wonderful polarity between richly differentiated light-dark perceptions and color perceptions that are so important in the three-dimensional representation of the forms, and also the geometric considerations about shadow form and position that touch on the activity of the senses of motion and balance.[62] The teacher can very informally point out such polarities.

*Fig. 130: The Shadow of a Cone (E. Wannert)*

*Fig. 131: The Shadow of a Cone on the Floor and Wall (E. Wannert)*

*Fig. 132: The Shadow of a Sphere on the Floor and Wall (E. Wannert)*

The class will likely have some students more strongly oriented mentally toward differentiated perceptual ability and others who are drawn more strongly to the geometric laws and principles. These varied abilities can be made fruitful for everyone. Any attempt to assign a value should be completely avoided.

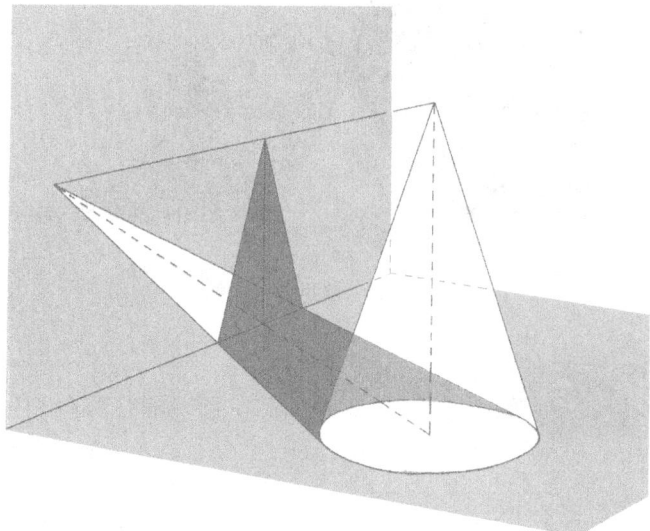

*Fig. 133: Construction of a Conic Shadow on the Floor and Wall*

As an intermediate step before the central projection on curved or bent surfaces recommended by Steiner, one can let a cone or ball shadow appear on two surfaces that are standing perpendicular to one another; for example, onto one level and one perpendicular standing image area. A portion of the form's shadow appears on each surface. With the ball two elliptical shadow parts are created that are stretched out in different directions (Figs. 131–133).

However, should the shadow of a cone with parallel lighting be drawn upon a step, for example, one must clarify where the shadow point is to be found on the vertical plane as well as the horizontal plane. The construction drawing makes this clear (Fig. 134). Again, the shadow point can be freely chosen.

Central projection on bent surfaces, as suggested by Steiner, comes about when the shadow-forming surface, the image area (white cardboard) is bent to a cylinder or when a second, larger styrofoam ball is used (Figs. 135–136). Working with these shadow forms is preparatory practice for the penetration of forms as they are taught in the seventh grade. In the sixth grade one can leave it with observations of this kind. A more thorough study of shadows of forms with a compass and ruler appears to me to be impractical considering the time limitations.

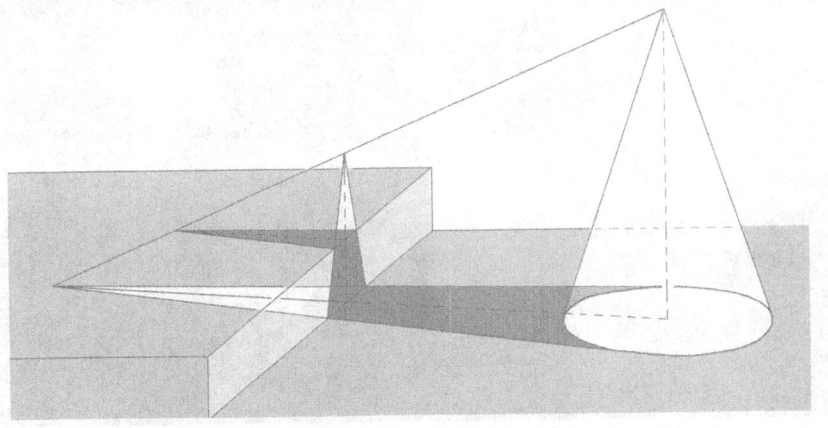

*Fig. 134: Construction of a Conic Shadow on a Step*

*Fig. 135: The Shadow of a Ball on a Sphere (E. Wannert)*

*Fig. 136: The Shadow of a Ball on a Cylinder (E. Wannert)*

# Geometry and Mineralogy

In the lectures Steiner gave at the founding of the first Waldorf school, he explained about the sixth grade curriculum: "Now the time is also here where we, the users of geometric forms, can transition to the mineral kingdom. We examine the mineral kingdom within a continuing relationship to the physical, that we also use with human beings, as I have already said; refraction—the lens for the eye; physically and chemically."[62] Space allows for only a few remarks here:

A formative force finds its expression in the crystal forms, as it is characteristic of the organism at a higher level. Its form must not be impressed on it from the outside, as is the case with all synthetically produced objects. Rather, with crystals the formative force lies within the material itself. This appears not to be widely understood by many people because they often look at crystals as synthetic and are amazed that they are found in nature. Naturally, one must realize that every higher organism subordinates its form in the time of a change of form, and during the exchange of material (metabolism) the primacy of the formative principles attest to the material. Crystals grow by layering. They have no exchange with their environment that could be described as metabolism. One principle of form rules a crystal from the beginning, from the first moment of its formation. This principle of form is an expression of the material that builds the crystal and the conditions (pressure, temperature) under which it is formed. In actuality, one and the same material can form into very different crystals, as in pyrite crystals mentioned previously. In crystals nature makes use of geometry in the simplest ways. The students can experience this if they are already familiar with the Platonic Solids and, if possible, constructed them.

The teacher can point out to the students that not all the Platonic Solids can be found in crystals. The icosahedron and the pentagonal dodecahedron are not. Nevertheless, other forms in nature can also be found for these solids. In the upper grades it will be possible to go deeper into the laws and principles governing crystals.[63] The fundamental principle will be found that says only certain surface angles appear in crystals. If there is a colleague at the school with the necessary equipment, the protractor for surface angles can be introduced and used on different crystals. If needed, the measuring is also possible with the help of two crossed pieces of paper.

It must remain open whether a deeper study of the principles governing crystals is possible in the sixth grade. It would be very interesting to hear about your experiences.[64]

# Outlook

The topics discussed in this book include study materials and projects that can be managed by students after age twelve and can be of help in developing their inner power of thinking and imagination in an age-appropriate manner. It is understood that not all students are equally capable of delving into spatial relationships as they are treated in geometry. Therefore, an inner differentiation in demands is essential. At some point, every student should have the feeling: I understand this! The many beautiful figures that can be achieved from geometry help to develop joy in the subject. This is especially true in this age of media where outside content threatens to flood one's consciousness; this certainty takes on extraordinary educational significance. There is material to be learned solely from inner observation and activity that is not subjective like the countless fantastical images that are absorbed. In the same way, all moral impulses must be gained from the inside if they are to be really autonomously grasped. "People do this or that" is not a fully responsible action, but rather, a certain understanding of a situation stemming from one's own inner being and intuition born of a moral imagination are human actions in the truest sense of the word.

How far an individual teacher goes in studying these subjects depends upon the students and the teacher's own abilities and time. There is no special mathematical prowess required. The students especially like to learn from a teacher who can, him or herself, still be amazed and find joy in a solution to a problem or a beautiful geometric figure. The expert, to whom everything is self-explanatory and who has no more questions, will generally find less rapport with the students than the teacher who is still learning.

Some teachers may perhaps stretch out *The First Steps in Proven Geometry* into the seventh grade, especially the suggested assignments. Every teacher must choose for him/herself. The next volume will go deeper into the topics that have been discussed here as well as help set a new course for the study of proportion and the beginnings of descriptive geometry.

The study of proportion leads into an area that can be perceived as "musical," and the spatial conceptions are permeated with a kind of study in harmony. In figurative geometry movement is introduced into the geometry to a much greater degree than would be possible without it.

The author appreciates input drawn from your own personal experiences that could be taken into consideration for the next printing of this series.

# ENDNOTES

1. See Ernst Schuberth, *Mathematics Lessons Grades 4 and 5*, New York: AWSNA Publications 2000.
2. See Ernst Schuberth, *Mathematics Lessons for the Sixth Grade*, New York: AWSNA Publications 2002.
3. This is related to the Greek god of wisdom—Jupiter. It may be interesting that Jupiter, the planet, needs about 12 years to come back to the place where it was when a person was born.
4. See Hermann von Baravalle's *Geometrie als Sprache* der Formen, Stuttgart: Verlag Freies Gestesleben, 1980. See also *Patterns in Space* by Colonel Robert S. Beard, Palo Alto, CA: Creative Publications, Inc. 1973.
5. Louis Locher-Ernst, *Space and Counterspace,* New York: AWSNA Publications, 2001.
6. See endnote #1.
7. Ibid.
8. Even if you have done this as recommended in Volume 2, you should repeat it here.
9. *Education for Adolescents*, GA 302, New York: SteinerBooks, 1996. Rudolf Steiner; 3rd Lecture given on June 14, 1921. See also *Waldorf Journal Project #12* for research on the importance of sleep.
10. See pp, 39ff.
11. It is at the same time Rule 1 from the first book, *Elements*, by Euclid.
12. See information about the complete solution of a geometric problem in Volume 2, p. 63. The steps given from the Greek are:

    Protasis: General problem in constructing an equilateral triangle
    Ekthesis: The specific case here requires a side length of 6 cm
    Analysis: Finding the solution. I know where the sides should join and that the second end point must lie on a circle.
    Apagoge: The construction arises from the characteristics of the circle.
    Kataskeuae: Doing the constructions
    Apodeixis: Proof that the construction fulfills the requirements
    Diorismos: Limitations

13. Naturally, it can be that the students have already suggested this complete solution. Then at this point, one would discuss the simplified solution with small arcs.
14. See Ernst Schuberth, *Social Education through Mathematics Lessons* in *Waldorf Science Newsletter, Vol. 8, No. 12, Autumn 2000*, Editor, David Mitchell, available on www.waldorflibrary.org.
15. There is a wonderful presentation (in German) by Martin Wagenschein on the six-pointed star in *Verstehen lehren: genetisch – sokratisch – exemplarisch*, Weinheim/Basel: Beltz Verlag, 1968/1999.
16. See Volumes 1 and 2 of this series. Volume 1, *Form Drawing Grades One through Four,* Laura Embrey-Stine, Ernst Schuberth, Fair Oaks, CA: Rudolf Steiner College Press; Volume 2, *Geometry Lessons in the Waldorf Grades 4 and 5*, Ernst Schuberth, New York: AWSNA Publications.
17. In some cases it is not very easy to explain which triangle should be held closer and which is to be held further away.
18. In terms of projective geometry, I tend towards indicating as many full lines as possible, not just segments. Through this the students should be given an outlook beyond the limited triangle into the whole plane.
19. See Volume 2, p. 73.
20. See Rudolf Steiner, quoted from *Rudolf Steiner zur Mathematik*, compiled by U. Kilthau and G. Schrader, Stuttgart: 1994.
21. See "Angles on Parallels" in Volume 2, p. 81.
22. On pp. 22 and 24ff we designated the number of angles with $A$ in order to point out the Angles. Now, we will assume use of the more commonly used designations—$n$ means any number.
23. See p. 24.
24. See Volume 2, p. 69f.
25. *Cathetus* is derived from the Greek κατηετοσ and means "perpendicular." The Greek verb *kathienai* means to "let down." Thus a *cathetus* (kathete) is a straight line falling perpendicular on another straight line or surface. *Hypotenuse* comes from the L.L. *hypotenusa* meaning "stretching under" (the right angle).
26. Thales von Milet lived from approx. 625 to 545 BC. He was known as one of the seven wise men of Greece and was said to have been one of Pythagoras's teachers. He taught that water was the origin of all things. In this sense, Goethe had him speak in the classic "Walpurgisnacht" (*Faust II*, Act II). See also Thales in: Rudolf Steiner, *The Riddles of Philosophy*, GA 18, New York: Anthroposophical Press, 1973, p. 51.
27. Those knowledgeable about projective geometry see that in the Theorem of Thales I, the creation of cone sections from projective aspects appears in an age-appropriate preliminary form.

28. See also Gerhard Ott, Geometrie für Klassenlehrer der 6., 7., und 8. Klassen, Private publication Vienna o.J. (1977), and Walther Lietzmann, Der pythagoreische Lehrsatz: mit einem Ausblick auf das Fermatsche Problem, Stuttgart: 1953.
29. Rudolf Steiner, *Erziehungskunst, Seminarbescprechungen und Lehrplanvortraege* (*Mathematics as an Introduction into Spiritual Knowledge*), GA 295, 10th Lecture, Dornach: 1984.
30. Compare Louis Locher-Ernst, *Mathematik als Vorschule zur Geisterkenntnis*, Dornach: 1973. This is the so-called Indian Proof.
31. The first formulation of the Pythagorean Theorem presented in the fifth grade applies only to the isosceles–right triangle. For information about the life of Pythagoras, see Ernst Bindel's *Pythagoras*, Stuttgart: 1962, especially the chapter "Leben und Wirksamkeit des Pythagoras," p. 36.
32. Rudolf Steiner, *The Study of Man*, GA 293, 14th Lecture, London: Rudolf Steiner Press, 1975.
33. What is meant is The Law of Cosines. It can be understood as the generalized law of Pythagoras. (See *Rudolf Steiner zur Mathematik*, compiled by U. Kilthau and G. Schrader, Stuttgart: 1994, pp. 124, 126, 163).
34. The relationship to the writing form for mathematical powers, which is here in the background, should only be indicated here.
35. Quoted from Ernst Bindel, *Pythagoras*, at the place sited (notation 31), p. 128.
36. Ibid.
37. In later grades we will show that this equality applies to all similar surfaces.
38. For more information on carrying over angles see Volume 2, p. 78.
39. As a reminder: In the English language, there is not yet a distinction between distance *from*, *Abstand,* and *distances apart, Entfernung.*
40. See Albrecht Beutelspacher, *Mathematik zum Anfassen. Bilder, Beschreibungen, mathematischer Hintergrund*, published by Foerderverein zur Schaffung eines Mathematikmuseums in Giessen, Inc., Giessen: 1998; Peter Baptist, *Die Entwicklung der neueren Dreiecksgeometrie*, BI, 1992, and the interesting article found on the Internet at http://did.mat.uni-bayreuth.de/~matthias/geometrieids/minimum/index2.html (8-9-2000).
41. See also the construction of angle bisectors in the fifth grade.
42. The last two cases should only be given as an assignment if the teacher is very clear about the relationships in the projective planes.
43. See Michael Toepell, *Platonische Koerper*. In MU 37/1991, Issue 4.

44. Rudolf Steiner; quoted from *Rudolf Steiner zur Mathematik*, compiled by U.Kilthau and G. Schrader; Stuttgart: 1996, p. 116.
45. See also the writing by Ernst Buehler in: Paul Adam and Arnold Wyss, *Platonische und Archimedische Koerper, ihre Sternformen und polaren Gebilde*, Stuttgart: 1994, pp. 16–17.
46. Paul Schatz, *Rhythmusforschung und Technik*, Stuttgart: 1998. See also http://www.paul-schatz.ch.
47. It says that in projective (real three-dimensional) space, for every figure and every theorem there is a polar (dual) figure and theorem in which the points and planes are alternately interchanged while the line segments in their center positions remain. Locher-Ernst, Louis. *Space and Counterspace*, New York: AWSNA Publications, 1988.
48. The question of how many solids fit the different definitions of "regular" cannot be pursued here, but it does lead to extraordinarily interesting insights.
49. It would be different if one made the edges going in and out every other one.
50. Ernst Schuberth, *Der Geometrieunterricht an Waldorfschulen*, Volume 2, p. 69f.
51. Walter Kraul, http://www.spielzeug-kraul.de.
52. The teacher should also know that the pentagonal dodecahedron and the icosahedron with regular surfaces cannot appear in crystals. Pyrite does form pentagonal dodecahedron forms that are five-sided but not regular.
53. See Rudolf Steiner, *Discussions with Teachers*, in the place cited (note 29), London: Rudolf Steiner Press, 1967.
54. Rudolf Steiner, *The Essentials of Education*, GA 301, London: Rudolf Steiner Press, 1968.
55. Rudolf Steiner, *Conferences with the Teachers of the Waldorf School 1919 to 1924*, Volume I, GA 300a, Dornach 1995, p. 264 and following.
56. Caroline von Heydebrand, *Lesson Plans for the Free Waldorf School,* translated by Eileen Hutchins, London: Steiner Schools Fellowship, 1996.
57. For the upper grades one can read the compilation *Rudolf Steiner zur Mathematik Band I und II*; compiled by U. Kilthau and G. Schrader; published by the Paedagogischen Forschungsstelle at the Bund der Freien Waldorfschulen; Stuttgart: 1994.
58. See Volume 2, p. 39ff.
59. If the sun is not shining, or if you wish to follow Rudolf Steiner's suggestion about Zentralprojection (central projection), you may use candles or small light bulbs with batteries. For teacher's experiments for the class, small halogen bulbs are recommended. The physics teacher in the upper grades would probably have them on hand.

60. See also the practice for freehand geometry in Volume 2, p. 13. Steiner recommended a conceptual treatment of the subject in the eighth grade.
61. See Ernst Schuberth, *Teaching Mathematics for First and Second Grades in Waldorf Schools*, Fair Oaks, CA: Rudolf Steiner College Press, 1999.
62. Rudolf Steiner, *Practical Advice to Teachers*, GA 294, p. 139.
63. See Hans-Ulrich Schmutz, *Erdkunde in der 9. bis 12. Klasse an Waldorfschulen*, Stuttgart: 2001. See also Renatus Ziegler, *Morphologie von Kristallformen und symmetrischen Polyedern: Kristall- und Polyedergeometrie im Lichte von Symmetrielehre und projektiver Geometrie*, Dornach: 1998.
64. Renatus Ziegler, *Morphologie von Kristallformen und symmetrischen Polyedern: Kristall- und Polyedergeometrie im Lichte von Symmetrielehre und projektiver Geometrie*, Dornach: 1998.

# BIBLIOGRAPHY

Adam, Paul and Arnold Wyss. *Platonische und Archimedische Korper, ihre Sternformen* und *polaren Gebilde,* Stuttgart: Verlag Freies Gestesleben, 1994.

Baptist, Peter. Die *Entwicklung der neueren Dreiecksgeometrie,* BI - Wissenschaftsverlag, Mannheim-Leipzig-Wien-Zürich, 1992.

Baravalle, Hermann von. *Geometrie als Sprache der Formen,* Stuttgart: Verlag Freies Gestesleben, 1980. See also *Patterns in Space* by Colonel Robert S. Beard, Palo Alto, CA: Creative Publications, Inc., 1973.

Beutelspacher, Albrecht (editor). *Mathematik zum Anfassen,* Giessen: See: http://www.wissenschaft-im-dialog.de/fileadmin/redakteure/bilder/wissenschaftssommer/2008/Infobrosch%C3%BCre_Mathematik_zum_Anfassen.pdf, 1998.

Bindel, Ernst. *Pythagoras,* Stuttgart: Verlag Freies Gestesleben, 1962.

Heydebrand, Caroline, von. *Lesson Plans for the Free Waldorf School,* translated by Eileen Hutchins, London: Steiner Schools Fellowship, 1996.

Kratz, Johannes. *Geometrie I,* Munich: Bayer. Schulbuchverlag, 1989.

Lietzmann, Walther. Der *Pythagoreische Lehrsatz: mit einem Ausblick auf das Fermatsche Problem,* Leipzig: Teubner, 1965.

Locher-Ernst, Louis. *Mathematik als Vorschule* zur *Geist-Erkenntnis,* Stuttgart: Verlag Freies Gestesleben, 1973.

—————. *Space and Counterspace,* New York: AWSNA Publications, 2001.

Ott, Gerhard. *Geometrie fur Klassenlehrer der 6., 7. und 8. Klassen,* Privatdruck, Wien (privately published), see: http://www.waldorf-mauer.at/baust~3.html, 1977.

Schatz, Paul. *Rhythmusforschung und Technik,* Stuttgart: Verlag Freies Gestesleben, 1998.

Schmutz, Hans-Ulrich. *Erdkunde in der 9. bis 12. Klasse an Waldorfschulen,* Stuttgart: Verlag Freies Gestesleben, 2001.

Schuberth, Ernst. "Mathematikunterricht an Waldorfschulen und soziale Bildung," in *ZDM-Zentralblatt fur Didaktik* der *Mathematik* 87/6, Karlsruhe: See: ZDM – *The International Journal on Mathematics Education,* Springer, 1987.

—————. *Teaching Mathematics for First and Second Grades in Waldorf Schools. Math Curriculum, Basic Concepts, and Their Developmental Foundation,* Fair Oaks, CA: Rudolf Steiner College Press, 1999.

—————, and Laura Embrey-Stine. *Form Drawing Grades 1–4,* Fair Oaks, CA: Rudolf Steiner College Press, 1999.

Steiner, Rudolf. *Discussions with Teachers,* GA 295, New York: SteinerBooks, 1997.

—————. *Education for Adolescents,* GA 302, New York: SteinerBooks, 1996.

—————. *Faculty Meetings with Rudolf Steiner, 1919–1924,* GA 300a-c, New York: SteinerBooks, 1998.

—————. *The Foundations of Human Experience* (formerly called *The Study of Man*), GA 293, New York: SteinerBooks, 1996.

—————. *Practical Advice to Teachers,* GA 294, New York: SteinerBooks, 2000.

—————. *The Renewal of Education,* GA 301, New York: SteinerBooks, 2001.

———————. *The Riddles of Philosophy, GA 18*, New York: Anthroposophic Press (SteinerBooks), 1973.

———————. *Rudolf Steiner zur Mathematik. Eine Sammlung von Zitaten aus Gesamtwerk. [Quotations by Rudolf Steiner on the Teaching of Mathematics]*, Zusammengestellt von Ursula Kilthau und Georg Schrader. Herausgegeben von der Pädagogischen Forschungstelle beim Bund der Freien Waldorfschule, Stuttgart: 1994.

Toepell, Martin. *Platonische Körper*, in MU37, Seelze/Velber: Erhard Friedrich Verlag GmbH, 1991.

Tropfke, Johannes. *Geschichte der Elementarmathematik*, 4, Band, Berlin: de Gruyter, 1940.

Wagenschein, Martin. *Verstehen lehren: genetisch – sokratisch – exemplarisch*, mit einer Einführung von Hartmut von Hentig, Weinheim, Basel 1999, Vgl. auch: Christoph Räbiger, Martin Wagenschein Sechsstern – Eine Begegnung mit der Wahrheit, in *Neue Sammlung*, Jg. 30 (1990), pp. 25–32.

Ziegler, Renatus. *Geometrische Kristallmorphologie auf projektiver Grundlage. Zur Komplementarität von Morphologie und Strukturtheorie*, Dornach: Mathematisch-Astronomische Sektion am Goetheanum, 1999.

Made in the USA
Middletown, DE
23 September 2025